Time in Geographic Information Systems

Technical Issues in
Geographic Information Systems

Series Editors:

Donna J. Peuquet, The Pennsylvania State University
Duane F. Marble, The Ohio State University

Time in
Geographic Information Systems

Gail Langran

CRC Press
Taylor & Francis Group
Boca Raton London New York

CRC Press is an imprint of the
Taylor & Francis Group, an **informa** business
A TAYLOR & FRANCIS BOOK

First published 1992 by Taylor & Francis Ltd.

Published 2023 by CRC Press
Taylor & Francis Group
6000 Broken Sound Parkway NW, Suite 300
Boca Raton, FL 33487-2742

© 1992 by Taylor & Francis Group, LLC
CRC Press is an imprint of Taylor & Francis Group, an Informa business

No claim to original U.S. Government works

ISBN 13: 978-0-7484-0059-1 (pbk)

Visit the Taylor & Francis Web site at
http://www.taylorandfrancis.com

and the CRC Press Web site at
http://www.crcpress.com

British Library Cataloguing in Publication Data
Langran, Gail.
    Time in geographic information systems.
    — (Technical issues in GIS series)
    I. Title   II. Series
    910.901

Library of Congress Cataloguing in Publication Data
Langran, Gail.
    Time in geographic information systems / by Gail
Langran.
    p. cm. — (Technical topics in geographic information
systems)
    Includes bibliographical references and index.
    ISBN 0-7484-0003-6 (HC)
    1. Geographic information systems. 2. Time.
    I. Title.
    II. Series.
    G70.2.L32   1991
    910′.285--dc20   91-28810
                                              CIP

Cover design by Barking Dog Art

Phototypesetting by
RGM Associates, Southport, England.

Contents

Preface

A reasonable goal for geographic information systems is that they be capable of tracing changes in an area by storing historic and anticipated geographic data. The fact that today's digital information systems—both spatial and aspatial—tend not to trace data lineage, retain prior versions of information, or track past and future production milestones or activities means that users of digital information systems must go without what analog information systems provide in rudimentary form.

The analog systems that are being replaced by digital systems do tend to provide a historical view of the data: when an entity changes, a new card is stapled to an old one, a new line is added to an existing card to describe the change, or the old version of information is crossed out, yet still legible. Analog systems also tend to describe lineage. new entries are initialed or, at minimum, an individual's penmanship or graphic style provides a clue.

The question arises, then, of how to correct this shortcoming. This work provides a conceptual, logical, and physical basis for developing a temporal capability in a geographic information system. It introduces a conceptual model of geographic change that stores changes cumulatively as they occur, but does not store unchanged data in duplicate. The conceptual model is applied to common geographic data types as a demonstration.

The discussion reviews the literature of time in information processing at length and suggests ways to apply this research to serve geographic purposes. Such practical topics as clustering, quality control, and volume control are addressed.

Later chapters focus on the problem of responding to *ad hoc* queries to a spatiotemporal database. The discussion introduces a taxonomy of multidimensional access methods. Four distinct classes are selected from the taxonomy and a representative of each class is implemented on small spatiotemporal datasets.

Acknowledgments

This work was a highly personal undertaking, but the support and encouragement of others made the sustained effort possible and even pleasant. I have been unusually lucky. Nick Chrisman devoted many hours to reading about and listening to my ideas, and our debates were both loud and helpful. Other useful input, both intellectual and personal, was provided by Morgan Thomas, Babs Buttenfield, Tim Nyerges, and Dick Morrill.

I have been fortunate also to have sponsorship for this work. Funding permits a level of concentration that is not always possible to achieve otherwise, yet it is an exceptional organization that is willing to underwrite academic research. I am particularly indebted to Tom Baybrook of Intergraph Corporation, who championed this work; and to Dave Scott, also of Intergraph, who encouraged its transformation into a book.

1

Introduction

A brief flight of fancy

Imagine a time in the future when geographic information systems (GIS) are widely available and used. In that future, any of the following scenarios might occur.

Forest resource management

A government agency uses a GIS to manage natural resources on public lands. In addition to other topographic data, the GIS stores the complete history of all timber stands. Each history describes a progression of harvest cycles, during which the stand was planted, fertilized, thinned, sprayed, sold, harvested, burned, and rehabilitated.

The timber stand histories allow the agency to project long-term timber yields and to compare growth rates over time given the stand's aspect, soil, and silvicultural treatment. The agency's GIS also has software to forecast the spread of disease and the behavior of a forest fire given a set of external variables.

Urban and regional management

A county's administrators use a GIS to track the area's geodetic, property, census, and utility records. The GIS is updated incrementally using information contributed by surveyors, engineers, and assessors.

The entire community uses the GIS. The county planner gauges development rates along certain corridors, monitors compliance with county conservation goals, and evaluates long-term land-use changes as part of a zoning study. The district attorney compiles the cumulative investment records of a local land speculator to use in a case against him. A developer reviews the gradual encroachment of the river on riverside property. And the centennial committee is permitted access to the historical data to create an animated map of the city's evolution.

Research and development

A research organization uses a GIS for spatial analysis and for developing new GIS capabilities. Researchers study regional development, evaluate alternate locations for public facilities given demographic forecasts and local behavior patterns, measure the long-term effects of pollution given various climatic and economic scenarios, analyze accident and disease patterns, and assess land-use or demographic trends.

In support of this research, the GIS has several advanced capabilities. It facilitates the production of maps and animations of spatiotemporal distributions via easy-to-use commands. Access to the stored data is also easy, particularly when compared to the punch card and magnetic tape procedures that many still remember having used. Statistical software is integrated with GIS software, as are utilities that translate numerical models into computer instructions.

Electronic navigation chart

A cruise ship line has installed electronic navigation charts aboard its island-hopping vessels. The graphic display, generated in real time from a stored hydrographic database, resembles an animation of a ship crossing a nautical chart. Pilots use the electronic charts to compute speeds and distances, for collision avoidance, for planning, and to maintain the routine portions of the ship's log. Further, each ship's lounge can tune its television to the electronic chart, which is broadcast over a closed-circuit channel for passenger amusement.

Because chart currency is crucial to navigation, the electronic chart database receives daily updates in the form of change notices via electronic links. The system automatically journals the change notices and enters them into the database. The risk of automated updates is minimal because they supersede but do not delete the previous database version and because the disseminating agency has already subjected them to rigorous quality checks based on previous object versions and their change periodicities.

En route, the navigator notices a database error. He or she compiles a change notice, enters it in the database as a provisional amendment, and broadcasts it on a channel reserved for that purpose. Others entering the area also enter it in their databases provisionally, pending official sanction. Once again, the risk of doing so is minimal, since the provisional update does not overwrite the current database version.

Infrastructure management

A utility company maintains a temporal GIS that describes public works improvements, routine maintenance, and breakdowns. In addition to tracing work in progress, the database describes future plans, current schedules, and the history of completed work.

The historical database has proven particularly useful in detecting weak links in the utility system. For example, because the database tracks when streetlight bulbs are replaced, the utility company's planners can identify streetlights with unusually high burnout rates and service them—a preventative measure that ultimately saves money. The database's future tense is also useful in scheduling; in addition to planning for resource allocation, knowing both when and where maintenance will be needed helps to minimize the movement of equipment and optimize the usage of available personnel.

Transportation

A state department of transportation uses a temporal information system to monitor performance and maintenance of state roadways. Maintenance scheduling is handled similarly to a utility department; in this case, analysts provide the system with road surface life expectancies, and update it with 'pothole reports' and other such input so it can anticipate maintenance needs.

Among other things, the highway safety department uses the temporal database to track traffic accidents, not just by location but also by date, time, season, and lighting. The accident-tracking utility produces descriptive animations at many time-space scales (e.g. showing intersections, towns, counties, states and lasting days, weeks, months, or years).

Map and chart production

A map producer maintains a geographic database of all the versions of all the features appearing on all of the company's products. The database's future tense allows analysts to post scheduled changes as reports of them are received; however, the changes do not affect the current data until their effective dates, at which time they are 'rolled in' to the database. The database's past tense provides a full accounting of database states. This is particularly important to hydrographic chart production, since lawyers can respond quickly to lawsuits claiming damages due to chart error.

Discussion

The enviable GIS capabilities described in these scenarios bear only a nodding acquaintance with those presently available. Some current GIS shortcomings relate to bureaucratic problems, others to the unavailability or unreliability of data, and still others to the expense of appropriate hardware and software. However, some aspects of each scenario are impossible strictly because they can be performed only by a temporal GIS, which does not currently exist.

The lack of these capabilities is by no means due to a lack of desire for them. One has only to scan the proceedings from recent GIS conferences to discover separate and uncoordinated forays into the temporal domain. For example, Burrough *et al.* (1988) discuss the possibility of linking spatial process models with GIS. Samsel and Colten (1988) explore the use of historical land use information to determine possible sites of accumulated hazardous wastes. Fifield (1987) describes a land use model to forecast land uses based on the current state. Holloway (1988) describes the efforts of North Carolina to digitize its land records so title can be traced easily and so updates to cadastral maps are made daily. Johnson and Sideralis (1989) describe their efforts to unite many 'project datasets' collected over the years into a 'corporate GIS database'. In general, the tenor of GIS applications is shifting from one of *ad hoc* problem-solving and demonstration projects to one where the GIS is institutionalized and long term. For this shift to be complete, some treatment of change over time is essential.

A reasonable goal for geographic information systems (GIS) is that they be capable of tracing and analyzing changes in spatial information. An atemporal GIS describes only one data state. This means that historical states are essentially forgotten and the anticipated or forecast future cannot be treated. Because of this, an atemporal GIS obscures the processes that cause states to change from one to the next, making dynamics of the modelled world difficult to analyze or understand.

In contrast, a temporal GIS would trace the changing state of a study area, storing historic and anticipated geographic states. By storing temporal information on-line, a temporal GIS could respond to the following queries:

Where and when did change occur?
What types of change occurred?
What is the rate of change?
What is the periodicity of change?

Given access to these types of data, software might assess:

whether temporal patterns exist
what trends are apparent
what processes underlie the change.

Such assessments could form the basis for understanding the causes of change, leading to a better understanding of the processes at work in a region. In sum, an ability to work with time could herald a new stage of GIS (Table 1.1).

Table 1.1 Three stages of GIS capabilities.

Stage	Input	Analysis	Output
First	*ad hoc*, in-house digitization	none; stores and retrieves digitized maps	hardcopy; goal is to replicate existing products
Second	centralized data capture, data exchange	single-state analysis, static modelling	interactive softcopy graphics; successful replication of existing products
Third	incremental updates, dissemination of change data	multi-state analysis, dynamic or predictive modelling	animated graphics, multitemporal maps, new product designs

Functions of a temporal GIS

The fundamental functions of a temporal GIS are inventory, analysis, updates, quality control, scheduling, and display (Table 1.2). This listing blends the functions identified by others.

For example, Tomlinson *et al.* (1976) list management, data acquisition, input and storage, retrieval and analysis, information output, and information use as fundamental GIS subsystems. Clarke (1986) merges Tomlinson's management and information use functions into one but leaves the others intact. Chrisman (1983) adds quality control to the list. Crain and

Table 1.2 Major temporal GIS functions.

Inventory	Store a complete description of the study area, and account for changes in both the physical world and computer storage.
Analysis	Explain, exploit, or forecast the components contained by and the processes at work in a region.
Updates	Supersede outdated information with current information.
Quality control	Evaluate whether new data are logically consistent with previous versions and states.
Scheduling	Identify or anticipate threshold database states, which trigger predefined system responses.
Display	Generate a static or dynamic map, or a tabular summary, of temporal processes at work in a region.

MacDonald (1983) distinguish land management from land inventory and analysis. Rhind and Green (1988) restate the functions as data input, encoding, manipulation, retrieval, analysis, display, and management. And Guptill (1988) adds 'user interface' to his listing of functional components. Because no clear consensus exists concerning GIS functions, I have chosen to define functions according to their objectives rather than the procedures used to perform them. The text that follows briefly introduces the functions and Chapter 2 discusses present and potential capabilities of each.

Inventory

A critical temporal GIS function is to store the most complete possible description of a study area, including changes that occur in the living world and in the database. A temporal GIS should be able to supply the complete lineage of a single feature, the evolution of an area over time, and the state of a specified feature or area at a given moment.

Analysis

The ultimate goal of analysis is to understand the processes at work in a study area. Since atemporal databases neglect change, they necessarily neglect processes. Temporal databases should improve the analytical capabilities of GIS by including specific references to change.

Analysis may be statistical, where data are examined for trends, patterns, divergence from the norm, and cross-tabulation or autocorrelation. Alternately, analysis may compare the empirical stored data to theoretical models to see where they diverge, or develop numerical models that approximate the data's character.

If the presence of a process is known or its workings are understood, it may be possible to forecast future states of the study area. Forecasting generally follows one of two approaches. The hypothesized system at work in an area can be treated as a black box, and its next states extrapolated from the preceding progression of states and expected intervening events. Alternately, a model can be devised to characterize each variable's role in the system's metamorphosis.

Scheduling

A temporal GIS could provide support for a facilities management system whose functions include scheduling. In that case, predetermined thresholds (e.g. streetlight burnout rates or chart changes) could trigger the system to notify its managers of the need for action (e.g. streetlight servicing or a new chart edition). A more advanced scheduling system might venture forecasts of when thresholds will be reached, based on the rates at which the measures are incremented.

Display

Among the more rewarding pay-offs of a temporal GIS is the ability to illustrate real-time occurrences (e.g. for electronic navigation or feedback during data collection) or to respond to queries via animated maps. Such queries as, 'how has this area changed over time?' or 'how will this process affect (or be affected by) the area within which it occurs?' are clearly candidates for animated response. Static maps of temporal geographic information are also important for portability, annotation, and contemplation. Finally, aspatial temporal information can be described in tables (for example, the changes to a feature's attributes over time) and atemporal spatial information can be described via traditional map designs (for example, a map of the study area as of a given moment).

Updates

A common method of updating geographic data is to replace an old dataset (completely) with a new one. A more rational approach to updates, particularly as datasets grow and electronic communications are employed, is to supply updates as change notices to expired data. The Notice to Mariners system is a prime example of change-only updating. Notice to Mariners, disseminated weekly to users of paper hydrographic charts, supplies a date, a location, and a change, which the user then transcribes to the affected charts.

Because digital geographic databases have only recently become sufficiently popular to be considered 'products' in the same sense as paper maps, and because most such databases are still in their first generation (i.e., they have not yet been updated), little thought has yet been given to update procedures. Two issues are critical: the logical and physical form of the updates, and how to maintain the outdated data as a resource.

Quality control

The work of the Task Force on Digital Cartographic Data Standards (Morrison, 1988) spotlights the need to design descriptions of and controls upon data quality into digital databases. This work states that it should be possible to describe fully data lineage, positional and attribute accuracy, logical consistency, and completeness. Arguably, temporal information can play a role in preserving or describing many of these components of data quality.

Technical requirements of a temporal GIS

Five technical requirements will drive the development of a temporal GIS: a conceptual model of spatial change, treatment of aspatial attributes, data-processing logistics, a spatiotemporal data access method, and efficient algorithms to operate on the spatiotemporal data.

The conceptual model is the configuration of information as it will be represented to the computer. It defines the entities, attributes, and relationships to be portrayed; it also defines the operations to be performed and the constraints to be enforced. This model is roughly equivalent to the 'infological' model of Sundgren (1975; also, see Tsichritzis and Lochovsky, 1982), which relates to human concepts and real-world representation.

Until random access to storage media became possible, procedures for processing data in storage and those for processing data in main memory differed sharply. However, the distinction between program and database representation is now becoming somewhat blurred. As argued by Fagin *et al.* (1979), random access to storage media now permits us to devise general algorithms to treat data using common methods, regardless of their storage location. While my focus does lean toward database representation because of the expected bulk of spatiotemporal data, the ultimate goal is to develop a conceptually sound basis for representing spatiotemporal data in any digital environment.

Given a conceptual model of real-world spatiotemporality, it is still necessary to describe that conception to the computer somehow, i.e. perform the 'datalogical' modelling of Sundgren (1975). Data-handling logistics include global partitioning, error control, and updating the stored data. A scheme for accessing stored spatiotemporal data is also critical, since their potentially enormous volumes could otherwise cause a system to perform too poorly for practical use. The design of such an access scheme was among my primary study goals.

Subsequent discussions require a set of standard terminology. The components of geographic information exist or can be represented at three levels: entity, object, and symbol (Nyerges, 1980a; 1989). Nyerges' definitions, and those I use here, are in accord with those of the proposed US national standard (Morrison, 1988). 'Entity' describes phenomena in the real-world, 'object' describes its database representation, and 'symbol' is a graphic representation of the data. Each of the representational types can describe geographic information, but they do not necessarily describe geographic information. When an entity is geographic, its objects are comprised of both spatial and attribute descriptors.

Objectives and sequence of this book

The goal of this work is to suggest an approach for further development of temporal GIS. Of necessity, this encompasses philosophical, conceptual, and technical decisions, which the following chapters address in roughly that order.

Chapter 2 describes previous work that has sought to conceptualize, analyze, and represent spatiotemporality.

Chapter 3 introduces the construct of cartographic time, examines models of spatiotemporality, and defines a conceptual model to serve as a basis for digital representation.

Chapter 4 discusses how to implement the conceptual model on selected cartographic data types.

Chapter 5 discusses ways to handle the attribute information that is a critical element of any GIS, and which is particularly important to the conceptual model devised here.

Chapter 6 presents alternatives and trade-offs for organizing spatiotemporal information, and for controlling data volume and error in a temporal GIS.

Chapter 7 discusses the problems of accessing spatiotemporal data from storage. It examines existing multidimensional access methods and develops a taxonomy to describe them.

Chapter 8 selects four opposing access methods, implements each using small test data sets, then queries the implementations to evaluate their effectiveness.

Chapter 9 summarizes the knowns and unknowns of the temporal GIS approach presented here and offers my conclusions.

2

Background

Geography differs from geometry because in geography, space is indivisibly coupled with time (Parkes and Thrift, 1980). While perhaps an understatement, this observation highlights the importance of time to all geographic endeavours. The preceding chapter sketched a temporal GIS and listed its functions: inventory, analysis, updates, quality control, scheduling, and display. The first step in designing such a system is to understand what existing literature applies to each function. A variety of methods currently exist for treating spatial data in a temporal framework. Many disciplines, including geography, cartography, and information science, have produced relevant work.

This chapter reviews the literature of spatiotemporality in the context of the six GIS functions presented in Chapter 1. Before embarking on a literature survey, however, a more specific definition of temporal geographic information is needed. Parkes and Thrift are not alone in their sentiments regarding the importance of time to geography. Berry (1964) describes geographic data in a matrix comprised of times, places, and attributes. Bullock et al. (1974) and Haggett et al. (1977) also use a three-dimensional matrix to describe the relationships of times, locations, and activities (activity is an attribute in this context). Dangermond (1984) depicts the elements of a GIS as time, location, and attribute, although today's technology treats only location and attribute.

Sinton's (1978) framework for geographic data representation draws the clearest distinction between current and desirable temporal capabilities (Table 2.1). Sinton agrees with others that time, location, and attribute are the three components of geographic information. However, he states that conventional data forms do not measure all three components concurrently. Rather, one is fixed at a constant value, a second is controlled to a range of values, and only then can the third be measured on an interval or ratio scale. Sinton defines fix, control, and measure by example. In fact, these terms elude precise definition because the nature of their application depends on the data being represented. For this reason, I follow Sinton's course of definition by example.

11

Table 2.1 The representation of geographic data in various formats (extended from Sinton 1978).

	Fixed	Controlled	Measured
Soils data	Time	Attribute	Location
Topographic map	Time	Attribute	Location
Census data	Time	Location	Attribute
Raster imagery	Time	Location	Attribute
Weather station reports	Location	Time	Attribute
Flood tables	Location	Time	Attribute
Tide tables	Location	Attribute	Time
Airline schedules	Location	Attribute	Time
Moving objects	Attribute	Location	Time

Most mapped data fix time. Data that fix location rather than time include time-sequenced economic, labor, or health statistics for specific locations; weather maps that measure minima and maxima over a given period at established stations, and tide tables that describe when high and low tide occurs at established locations. A greater challenge is to fix neither location nor time; to describe the path of a moving object would achieve that goal by fixing attribute (i.e. the object's identity), but the true challenge lies in not fixing any of the three components of geographic data. Describing changes in land cover, census tracts, and political boundaries meets this qualification by controlling one or two components and measuring the other(s).

Sinton's framework is important because representational constraints limit the topics and methods of geographic research. To create a temporal GIS is to dislodge the time component from its fixed position and thus free this log-jam in geographic inquiry. The sections that follow discuss the functions of a GIS with respect to these goals and constraints.

Inventory

The purpose of inventory is to enumerate and describe the critical components of a study area. Maps, which provide a systematic and compact archive for spatial facts, are fundamental to geographic inventory. Digital databases are gradually replacing paper maps as inventory tools because they permit facts to be stored irrespective of scale and symbol constraints, and because they facilitate many analytical procedures.

For this reason, the methods of inventory are closely associated with those of databases. Geographic databases rely on somewhat specialized methods because their spatial contents require it. Spatial data include ordered, interconnected, variable-length components (e.g. the vertices of lines and the

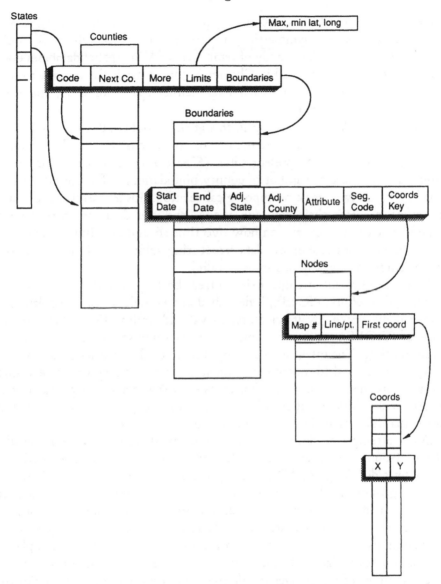

Figure 2.1 A data structure for temporal county boundaries designed by Basoglu and Morrison (1978).

lines that bound polygons) that are difficult to treat using database procedures developed for business and other aspatial applications. The spatial data management literature is vast; Nyerges (1980), Peuquet (1984), Drinnan (1985), Smith *et al.* (1987), Rhind and Green (1988), and Frank (1988) address a cross-section of current spatial database issues.

Databases are seldom both spatial and temporal. Apparently, even database designers feel compelled to fix either space or time because they are not conceptually free of symmetrical report formats, to which Sinton's constraints are tied. A large body of temporal work in aspatial databases does exist and is reviewed in detail in Chapter 5, but of particular interest to this discussion is database research that fixes neither space nor time, and thus foreshadows a temporal GIS.

One early work is outstanding in this area. Basoglu and Morrison (1978) devised a hierarchical data structure to store changes in US county boundaries from early to present times (Figure 2.1). The data structure is designed to produce a snapshot of county boundaries as of a requested date. It is lacking in two areas: no topological relationships are described because the effort predates their common usage, and the data structure cannot respond efficiently to queries on how two time slices differ. But while this is not a definitive spatiotemporal data structuring solution, it is an impressive first attempt to represent spatiotemporality.

A few more recent attempts exist to treat both space and time in a single database. One effort, PROBE, is described as a knowledge-oriented database management system by its creators, Dayal and Smith (1986). One of the goals of PROBE is to handle dimensional concepts so data are oriented in space and time. The intent is to develop dimensional semantics to augment a query language and to incorporate data structures that accommodate spatial and temporal needs. This research is in its early stages and is ongoing. Initial reports make little mention of how to treat geographic change at the conceptual level or of how to define the data to be stored.

Other attempts to develop a spatiotemporal database (e.g. Katz *et al.*, 1986a; 1986b; Beech and Mahbod, 1988) are geared to the needs of computer-aided design (CAD). The engineering design process is an iterative one; the goal of this development is to permit a designer to view all versions of a design and, if necessary, backtrack to a previous one. The Katz approach is representative of CAD strategies. Given a database of interconnected mechanical parts, information is organized in three planes: the version plane, the configuration plane, and the equivalence plane. The version plane stores all the versions of a part over its lifetime and permits the user to designate versions as active, superseded, or alternatives of one another. The configuration plane combines versions of different parts to form composites, e.g. to create new parts from component parts. The equivalence plane correlates the different versions of parts in different configurations, so separate versions of a part that is used in another part are recognizable as such.

While the Katz system does incorporate space and time, it is not clear whether a design so totally geared to mechanical engineering needs can also meet geographic needs effectively. For example, the flexibility of the version plane in designating alternate versions may be excessive (and therefore excessively expensive) for geographic purposes. Enabling different geographic configurations (for example, complex entities or maps) to contain different, but correlated, versions of their components is an interesting capability, but one that is perhaps treated too ornately here for geographic needs. Finally, Katz advances no conceptual model of change.

In sum, no blueprint exists for a temporal GIS database, although a great deal of literature applies to spatial or temporal database design. Spatial and temporal database techniques can, theoretically, combine to create a spatiotemporal database, but a conceptual problem remains: how to represent spatial change to the computer and identify precisely what objects will be stored.

Analysis

Inventory is not always an end in itself. A tacit goal is to improve our understanding of the world's spatial interactions and processes by examining sequences of events in a geographic context. Early works on geographic progressions include those of Davis (1899), who developed a theory concerning the life cycle of erosion, and Whittlesey (1945), who studied the sequent occupancy of human societies in ecological settings. But models of diffusion were among the first systematic approaches to studying the effects of time and space upon human affairs. Hagerstrand's seminal work in this area (1952), which studied the acceptance rate of Swedish farm subsidies, stimulated further study of the diffusion of other phenomena and ideas.

Diffusion models are useful for studying geographic processes that obey local laws. Tobler (1979) developed a more formal version of this continuous model based on the Game of Life (Gardner, 1970), which he calls Cellular Geography. Couclelis (1985) suggests the use of cellular automata (Codd, 1968) to break away from the grid and thereby simulate spatial interactions and multi-scaled processes more realistically. An intriguing aspect of diffusion methods is that they fix neither time nor location, although both are controlled (time in uniform snapshots of the study area and location in grid cells).

Hagerstrand also introduced Time Geography (1970), a less quantitative but equally evocative method of studying the effects of time and space on human affairs. Time Geography is best known for its spatiotemporal diagrams (Thrift, 1977, supplies a graphic overview). These three-dimensional 'aquariums' and 'prisms' identify the constraints on human decisions via the physical limits of activity, interaction, and behavior. According to Pred (1977), because many decisions are impulsive or intuitive, constraining the phase space is a valid way to examine potential outcomes.

Time geography's techniques and tools permit us to view geographic information without fixing or controlling time or location, although generally at the cost of fixing attributes (into object identities).

Others have applied statistical and numerical methods to the task of studying geographic processes, including spatiotemporal autocorrelation (e.g. Bennett, 1979; Cliff and Ord, 1981) and numerical modelling. Many such methods are distinguished by treatments of space and time that are incomplete or nonexistent. Some models use a single constant to express the dominant character of a region, implying that the effect of spatial heterogeneity is inconsequential.[1]

Some modellers do strive to represent spatial or temporal heterogeneity. Samson and Small (1984) model acid rain deposition rates to vary seasonally because of differences in vegetation cover—i.e. they represent temporal variations in a spatially homogenous region. A second acid rain deposition model (Carmichael and Peters, 1984) overlays a coarse grid on the study area, placing a single land cover value in each cell. Box (1981), in an ambitious attempt to predict world vegetation from climate alone, compiled a temperature and precipitation database of 1237 sites worldwide, computed a vegetation model at each site, then estimated vegetation worldwide. However, to measure spatial heterogeneity, Box controlled temporal heterogeneity by describing climates via minima and maxima.

The limited availability of spatiotemporal data ultimately affects geographic analysis. As stated by Morrill (1963), we can develop theories that explain processes by studying real patterns, then try to reproduce the actual patterns by applying the theories. But to study real patterns requires real data and seldom does a researcher have access to data describing more than one spatial time slice. When temporal data do exist, they generally fix or control location or attributes. It is not surprising, then, that most spatial theory is static and prescribes an optimum equilibrium state (Isard, 1970; Morrill, 1977).

Cliff *et al.* (1979) argue that lack of data affects the subjects that are studied—hence the concentration of studies on labor markets, population, and infectious disease. Since the introduction of digital cartographic methods, however, data stores are gradually growing. An imporant and

---

[1]Considerable controversy over this limitation exists within the area of predator-prey models alone. For example, Harrison (1979) and Rosenblat (1980) state that environmental heterogeneity and fluctuation have minimal long-term effect on predator-prey relationships and are rightly excluded from their models. Conversely, Engstrom-Heg (1978), Rosenzweig (1971), Comins and Hassell (1976), and Ludwig *et al.* (1979) argue that the environment has a causative influence on the predator-prey relationship, but limit their models to single-parameter stochastic descriptions of the environment.

highly spatial form of analysis operates on a categorical coverage (Chrisman, 1982), which measures attributes that exhaustively cover a region (e.g. soils or land-use data). Aside from air photos and satellite images, very little temporal coverage data exist, and when they do, the collection intervals may not suit the needs of a researcher. Analyzing temporal coverages is not entirely straightforward. The remote-sensing literature is replete with examples of change-detection procedures, which generally revolve around subtracting one state from the other to produce a subset of pixels that differ between the two states (see Colwell, 1983 for an overview of methods).

The problem of detecting changes between two snapshots lies in determining whether the differences are change, or whether they are errors in processing and registering the two states. A second problem is that the amount of data that must be processed is constant, regardless of the extent of change that has occurred. Overlay methods are also used by GIS researchers to operate on spatial data, with similarly inscrutable results. Steinitz *et al.* (1976) describe an example of early change-detection procedures. A 1912 design competition in the city of Dusseldorf resulted in the publication of *Sonder-Katalog für die Gruppe Stadtebau der Stadteausstellung zu Dusseldorf 1912*. Included in this volume were five depictions of Dusseldorf's development between 1874 and 1912. All five maps were drawn to the same scale so they could be overlaid and compared. Chrisman (1982) and Ross (1985) provide more recent examples of vector-based land-use change detection.

At minimum, a temporal GIS will facilitate using current analytic methods because data availability will improve, temporal depth will increase, and databases will ease the difficulties of working with large datasets. It is also reasonable to hope that the novelty of having truly spatiotemporal data to work with (i.e. where neither time nor space are fixed) will spur researchers to generate new ways of analyzing and understanding geographic processes.

Updates

The effect of an update is to substitute current for outdated information. Change is the impetus for updates, since change ultimately degrades the quality of geographic information. Grelot and Chambon (1984) and Armstrong (1988) discuss the types of spatial change that can occur, although neither suggests a treatment. However, traditional map update measures offer some guidance. Each government mapping agency uses a set of update procedures suited to its needs. In addition, incremental update procedures exist for long-term GIS maintenance.

Institutional updating methods

Perhaps the most ambitious update programs are Notice to Mariners (a nautical chart updating system), and Notice to Airmen (the aeronautical corollary). Both programs produce weekly reports of changes to charted and uncharted information in an attempt to support safe navigation. In both cases, navigators are legally required to transcribe the updates onto existing charts or ancillary materials. The paper trail of update notices bears some likeness to a temporal database because it describes incremental change and fixes only one of many attributes: object identity.

The US Geological Survey, whose topographic products are not intended for navigation, maintains maps by overprinting new information in a contrasting color. Overprints of updates highlight areas of change, although the visual effect of extensive overprinting can be quite poor. USGS guidelines state that consecutive overprints should use different colors to distinguish when each was current (Thompson, 1979).

Just as a topographic map series provides an inventory of an area's characteristics, overprints and sequent map editions provide a lasting archive of geographic change. However, identifying when change occurred or depicting intermediate changes via these methods alone is impossible. Even working with only the changes described on sequent editions is not straightforward. New editions are designed to replace old ones, not to be used side-by-side with them. A map's design can change between the production date of one edition and the next, meaning that scale, symbols, and colors do not necessarily match. In addition, the update cycle that determines when new editions are produced is generally based not on magnitude of change but on production pressures.

Exceptions to this rule include the US National Ocean Service (NOS) and the British Ordnance Survey. NOS updates a chart when the number of changes to that chart reaches a threshold value. Similarly, the Ordnance Survey has switched from an update system similar to the USGS. Their current methods are more akin to those of NOS because they record continuous and incremental change. Regional offices of the Ordnance Survey maintain versions of each map sheet in their jurisdiction, upon which surveyors mark changes as they occur. While this information is not published, it is available to customers in draft form. Originally, a new map edition was published when total changes exceeded 300 per sheet. Now, however, maps are published based on demand using the most recent available data.

Sinton's framework would place the update methods discussed here into different classes. When updates occur edition-by-edition, time within a map is fixed and time across editions is controlled to the publication dates of the editions. In contrast, incremental updates measure time, location, and most attributes, which makes these data an appropriate model for the contents of a temporal GIS.

Incremental updates to an atemporal GIS

Incremental update methods now exist for atemporal GIS that are equally well suited to temporal GIS. The difference between a temporal and an atemporal update is that the former supersedes the outdated information while the latter deletes or overwrites it.

As GIS databases mature, update procedures become more critical. Unfortunately, they are also quite expensive. Corson-Rikert (1987) cautions that maintaining current data can approach the cost and time involved in obtaining the original data and Goodchild (1982) estimates that, over the lifespan of a GIS, the effort required to maintain its data in current form is approximately 80% of the original capture effort.

Two types of GIS are evident in the literature. One type is created *ad hoc* to respond to a particular problem, then abandoned when the problem is solved. The second type is intended to be a permanent repository for an organization's data and the vehicle for many of its operational procedures. Recognition of this duality is evident in Dueker's (1979) distinction between routine and non-routine GIS operations, and Tomlinson and Boyle's (1981) 'inventory-related' and 'project-related' GIS. An *ad hoc* GIS may be dismantled following a project's completion and thus is not a candidate for update procedures unless the project examines spatial change or the data are revived for later use. In contrast, a GIS repository must be updated, preferably incrementally. Corson-Rikert (1987) discusses methods for updating a geographic database interactively, including comparing two time slices, adding known changes to an existing file, and interactively digitizing changes using graphic aids provided by the system.

Quality control

The role of quality control is to prevent errors from entering the database. The database literature uses the term 'data integrity' to describe this construct and includes it as the third component of a data model (Codd, 1981). Errors fall into several classes determined by cause. National standards of cartographic data quality are in the process of finalization (Morrison, 1988). To ensure that standards are met, each component of data quality must be guarded by established procedures. Temporal effects can make quality assessment more difficult, because two maps that are apparently inconsistent may actually be valid maps from different years (Chrisman, 1983). At present, distinguishing error from change requires human, not machine, decisions.

Origins of error

A first step in quality control is to consider how and where errors occur in the cartographic process. Walsh *et al.* (1987) identify two types of error: inherent and operational. Inherent error derives from the data source or is introduced during the capture process. Operational error is introduced through manipulation, misregistration, and misinterpretation. A final error type, 'use error', can occur when all operations are performed correctly but the user misinterprets the results (Beard, 1989).

Components of cartographic data quality

The National Committee for Digital Cartographic Data Standards defines five components of cartographic data quality: lineage, completeness, logical consistency, and positional and attribute accuracy (Moellering, 1987). As they exist now, these components apply largely to atemporal information. Chapter 6 discusses ways to extend them to data whose time component is not fixed.

Lineage describes source materials, data capture methods, and any transformations applied to the digital file. Essentially, all known dates that apply to source, capture, and update also comprise lineage. The dates when information was discovered in the physical world are preferred; if lacking, however, lineage dates can describe source publication dates if declared as such. Lineage also can be expressed through a 'quality overlay', which is a temporal composite.

Completeness implies that mapping rules are applied consistently to all data. Completeness encompasses selection criteria, geometric thresholds (e.g. minimum area or width), categorical classifications, and definitions. Completeness also describes whether data account for all space or if gaps occur.

Logical consistency addresses the fidelity of relationships described by the data structure. For example, the relational data model that is often used to treat geographic attributes has two general constraints: every table must have a unique identifier for the records it holds and foreign keys must match precisely across tables. The topological model used for most geographic data[2]

---

[2]Examples of topologically structured files include the US Geological Survey's Digital Line Graphs, The Defense Mapping Agency's Standard Linear Format, and the Census Bureau's TIGER system.

specifies that chains intersect at nodes, that cycles of chains and nodes be consistent around polygons, and that inner rings embed consistently in enclosing polygons.

Positional accuracy describes compliance with geodetic, survey, and production standards established for a given dataset, and must consider the quality of the product following all transformations applied to it in the capture process. Several possible ways to test positional accuracy exist: deductive estimation based on knowledge of error propagation rates; measurement of the final product to determine closure of traverse or residuals from adjustment; comparison to source; or comparison to a source of higher accuracy.

Attribute accuracy concerns the correct expression of object attributes in the file. For continuous-scale attributes, the same attribute accuracy measures can be used as for positional accuracy. For categorical attributes, estimates of accuracy can be based on deduction, independent samples, or polygon overlay.

Scheduling

A scheduling capability permits a database to predicate its actions upon predefined threshold states. Stonebraker (1986) describes how scheduling can be based either on present situation or on elapsed interval, a mechanism that he calls 'triggering' and that Segev and Shoshani (1987) call 'cause and effect'.

Many useful applications of scheduling exist in atemporal and temporal GIS, although the literature provides minimal discussion. An electronic chart could be instructed to increase the display scale as the vehicle or vessel approaches its destination and a map-production system could be instructed to schedule a new edition when the number of changes on a map surpasses a given number. Rhind *et al.* (1983) have attempted to use map revision data to anticipate where and how much change will occur, which could potentially permit scheduling procedures to forecast revision dates and allocate resources accordingly.

Triggers require an extended query language so that users can specify circumstances and the actions they should trigger. One problem of scheduling is the difficulty of conveying imprecise temporal meanings; 'as soon as possible' is common and useful terminology among humans but most computers would find this instruction unfathomable.[3] Triggers also

---

[3]Kahn and Gorry (1977) and De *et al.* (1985) elaborate on how the imprecision of certain important temporal expressions could be conveyed to the computer.

must correlate the times when events occur in the external world with internal state changes (Studer 1986). The constructs of world and database time, discussed at length in Chapter 3, are helpful, although not all researchers use such a complete treatment.

Triggers are inherently temporal, since the passage of time causes circumstances to change and trigger actions. But even triggers designed to be activated by spatial change require only that a threshold state be recognized and place no specific demands upon the longitudinal representation of that spatial change. To implement a scheduling capability in a temporal GIS requires a slightly different focus from the other functions discussed in this chapter. Triggers operate on data; they do not affect its expression and, more specifically, they do not affect what information components are fixed, controlled, or measured. For this reason, triggering methods relate more to algorithms than to data representation, and as such are distinct from the conceptual problems of the other GIS functions.

Display

Display is the way a GIS describes its contents and conclusions to its users. An atemporal GIS produces maps and tables to answer specific questions; likewise, a temporal GIS should be capable of producing maps and tables to answer temporal questions. This is not entirely straightforward. Fleming (1976) and Schneider *et al.* (1979) lament the lack of satisfactory methods to portray the dynamic patterns of moving objects on a two-dimensional map. Muehrcke (1978) notes that cartographers traditionally have maintained their composure in the face of a continually changing world by making static maps of relatively static phenomena, thereby shifting the burden of dealing with temporal phenomena to the map user. He lists four temporal constructs that require mapping techniques.

- Qualitative change, i.e. what changed?
- Quantitative change, i.e. what are the relative magnitudes of change?
- Composite change, where more than one time slice is mapped, e.g. paths, time sequences, diffusion, cycles.
- Space-time ratios to describe such space-time interactions as travel times, cognitive spaces, or rates of spatial change.

Grelot and Chambon (1984) suggest that GIS be capable of producing four different graphics that loosely correspond to Muehrcke's typology: maps that show differences from a base state (qualitative change), maps to describe present-tense data (atemporal), maps that detail before-after versions of information (composite change), and maps for special-purpose abstraction of information (conceivably, thematic maps of quantitative change or space-time ratios).

The mapping techniques that are available to describe these spatiotemporal constructs are relatively modest. Four major classes are evident.

- Time sequences, e.g. multiple editions or time series.
- Change data, e.g. text, graphic, or digital amendments to a base representation. Text change data include Notice to Mariners and Notice to Airmen. Graphic change data include the contrasting overprints used by the US Geological Survey and the change maps computed by subtracting two registered Landsat images of the same area. Digital change data could be delta or superseding values incorporated into file structures.
- Static maps with thematic symbols of a temporal theme, e.g. symbols depicting dates, rates, paths, or order of occurrence.
- Animations, where both space and time are scaled to depict the metamorphosis of a study area.

Sinton's framework is particularly germane to graphic representation. All the geographic data described in Table 2.1 fit within Sinton's framework of fixing, controlling, and measuring one each of the three information components. Not coincidentally, it is also possible to display all these data using two-dimensional graphics and tables, but time sequences and incremental updates breach Sinton's framework by fixing neither time, location, nor attribute. In effect, then, these temporal forms appear to add a 'dimension' to the information being represented and require similarly multidimensional display formats.

One such format is multiple maps of different time slices. A second is animation. Such queries as, 'how has this area changed over time?' or 'how well does this process interact with the area within which it occurs?' are clearly candidates for animated response.

Moellering (1980a) and Calkins (1984) review the techniques and applications of animated maps. Of interest here are 'temporal animations', i.e. rapid display of time series, not 'surface explorations' that permit a user to traverse, orbit, or circumnavigate a single-state display (e.g. Moellering, 1980b). The literature supplies examples of temporal animated maps used to show urban growth (Tobler, 1970b), the earth's seismicity (Levy *et al.*, 1970), satellite ground tracks (Turner, 1986), and automobile accidents (Moellering, 1973). Animated maps produced from meteorological satellite data are a common tool for weather reporting.

Maps play a crucial role in geographic research; improved access to temporal maps could lead to improved understanding of geographic processes. Borchert (1987, p. 388) remarks upon their importance.

Many time series of maps are in one sense statements of theories, in cartographic language, about geographic development processes, about the functioning and the past and future evolution of some global or regional system. Interpretations of the map patterns involve logical interpolation or extrapolation from mapped observations, in both space and time. Distinctively geographic models are also cartographic generalizations. As four-dimensional descriptions of the geographic evolution of resource and settlement systems, time series of maps are a fundamental element of geographical explanation.

Calkins and Dickinson (1987) suggest that the complexity of spatiotemporal data demands special tools to visualize the data, analytical processes, and results, and recommend adapting the data exploration methods of statistics to these tasks.[4] By this means, they argue, a researcher can identify space-time patterns within subsets of the data.

While powerful in their raw form, the full potential of static and dynamic temporal maps has not yet been explored systematically by cartographers, although Muehrcke (1978) and Tufte (1983) provide a sampling of static temporal mapping techniques. Calkins and Dickinson (1987) recommend exploiting colour to describe temporality, and describe one technique of allowing the colours of symbols to 'age' with time so the most current data is always the same colour while older data pass through a series of saturations coresponding to their currency dates. Aside from developing overall treatments of multidimensional information, spatiotemporal display designers must use softcopy media effectively (see Dobson, 1983) and express temporal and spatial scale effectively. Animations introduce more exotic concerns: animated symbols can flash, bubble, sparkle, throb, erupt, sink, rotate, shake, or explode by exploiting the third and fourth dimensions provided by spatial perspective and time.

Conclusions

This review suggests that, while researchers have not neglected spatiotemporal issues, the work that exists suggests only a bare sketch of a temporal GIS. For inventory, most research has focused on the admittedly difficult problems of spatial and temporal databases in isolation. Fully spatiotemporal attempts either lack a conceptual framework of geographic change or are geared to nongeographic applications altogether.

A large and diverse body of literature addresses spatiotemporal analysis.

---

[4]Work in data exploration includes that of Tukey (1977); McNeil (1977); Velleman and Hoaglin (1981); Hoaglin, Mosteller, and Tukey (1983); and Cleveland and McGill (1986).

However, virtually none of this work meets the criteria for spatiotemporality established here because most fix one of the information components. This neglect is quite understandable. Truly spatiotemporal data are rare, and those that do exist are difficult to treat because of their size and asymmetry. These factors suggest that revisiting existing spatiotemporal methods within a temporal GIS framework could lead to improved results and new methods.

Presently, the incremental update procedures used by some mapping organizations provide the best model of spatiotemporal data representation. Map and chart producers now use a variety of measures to maintain geographic information in current form. Most analog update methods have digital corollaries, and therefore do supply a guide to what works and what does not. In addition, GIS repositories are leading the way in developing incremental update procedures for digital data. All that remains is to retain the superseded data rather than to discard it. NOS is presently taking a major step in this direction with the Automated Nautical Charting System under development by Intergraph Corporation (see Langran, 1990).

Cartographic data quality has been defined and data producers can choose from a suite of quality-control methods. Quality control for a temporal GIS requires that we somehow extend these definitions and methods to data whose time component is not fixed.

Scheduling within a GIS framework is minimally treated in the literature. Nonetheless, the problems of developing a scheduling capability are somewhat distinct from those of other temporal GIS functions because they are relatively independent of the data's structure.

Because GISs communicate their results via maps and tables, developing temporal display formats is critical; because spatiotemporal data breach Sinton's framework, designing displays is not straightfoward. Maps comprise the more challenging problem; a limited set of formats for temporal maps do exist, but no conceptual or systematic work has been produced to guide a temporal GIS designer. Suitable tabular display formats also are required; animations and other thematic mapping techniques may be extensible to tables.

This work focuses on the conceptual and structural problems of representing geographic change. Thus, of the five functions of temporal GIS, its greatest impact will be on inventory, updates, and quality control. Improved analytical methods will follow improved data availability and manageability. Scheduling is so specific to applications and sufficiently independent of structural concerns that it is not addressed by this work. Display, while crucial, is a problem that merits full treatment and, for that reason, will be left to more dedicated study.

3

A conceptual model of cartographic time

The preceding chapters treat the functions and components of a temporal GIS in broad terms. This chapter narrows the focus and considers how to represent spatiotemporality to the computer. With that goal in mind, I frame a working construct—cartographic time—which is a scaled and generalized version of a more problematic phenomenon, time itself.

Defining cartographic time

Time, a phenomenon that can be perceived only by its effects, has stymied philosophers since its beginnings. Aristotle argued that time is the quantity of movement. By this reasoning, the hands of a clock and the segments of a road when travelled represent subdivisions of space that are linked together by time. Newton viewed time as a dimension that is separate from but similar to the spatial dimensions, in that all are containers for occurrences.

The turn of the century witnessed a popular fascination with the nature of the fourth dimension. During a period that began with Edwin Abbott's Flatland (1888) and lasted approximately forty years, the fourth dimension was presumed to be a spatial one—a sentiment that gave rise to non-Euclidean geometries, fostered a Scientific American essay contest (published as a collection by Manning, 1910), was exploited by spirtualists and charlatans, and inspired the cubist art of the period (Henderson, 1983). This flirtation with a fourth spatial dimension gradually faded after Einstein published his Theory of Special Relativity in 1905. By the 1920s, most scientists had accepted Einstein's premise that time is a fourth dimension that interacts with space (see Freeman and Sellons, 1971; Morris, 1984).

Our culture commonly views time as a line without endpoints that stretches infinitely into the past and the future, although compelling arguments exist for such alternate topologies as multiple parallel lines, tree structures, circularity, discreteness, and non-existence. (Rescher and Urquhart, 1971; Newton-Smith, 1980 provide excellent discussions of this topic; also see Ben-Ari et al., 1981). Among the interesting philosophical works on time are those of Rucker (1977 and 1984). Rucker argues that the passage of time is an illusion. Every human and physical object is a static pattern in a four-dimensional space-time block. Each moment, hence each configuration at each moment, exists permanently and concurrently in this block universe. Time only seems to pass because our memories work backwards. Thus, while we are ever present at every instant of our life, at each instant we remember only what has passed, not what is yet to come.

It is useful and interesting to examine the thoughts of philosophers concerning the nature of time and reality. But the goal of cartography is neither to discover nor to represent all of reality; if that were so, cartographers would have found alternate employment soon after air photographs became widely available. Rather, the goal of cartography is to distill from 'reality' its most important elements and represent them in a manner suited to their application. For this reason, cartographers can sidestep debates on what time is, and instead focus on how best to represent its effects. Isard's (1970, p. 8) comments are pivotal.

> I do not wish to think about time in any philosophical or metaphysical sense. I am searching for operational concepts of time—concepts that will lead to its direct measurement, to better theory, and to a richer set of models using time as a variable.

The questions, then, are what components of time are important to represent and how to represent them. Because of their representational focus, I call the answer to these questions 'cartographic time'. Cartographic time distills the characteristics of time that are essential for representing spatiotemporality in the most pragmatic and generic fashion possible. The definition that follows does not presume to be exhaustive; specific applications will doubtless add to or amend it. Nonetheless, the following basic components of cartographic time are important to address in a general-purpose temporal GIS.

Time as the fourth cartographic dimension

Cartography is an apparently two-dimensional activity, since its traditional products fit upon a sheet of paper. Through the use of symbols and tones, however, cartographers extend their domain to one or more additional dimensions, be they physical or thematic.

The attribute dimensions differ from the spatial dimensions (which is argued in Chapter 7). When added to cartography's two spatial dimensions, attributes describe a statistical surface, not a three-dimensional solid. Thus, to consider attributes a third dimension and time a fourth is not entirely accurate. Nonetheless, cartographic time will be considered as the fourth cartographic dimension.

Dimensions are inherently and unavoidably spatial constructs.[5] This and subsequent discussions develop temporal corollaries to spatial constructs through analogy. However, working with time alone is far easier than working with space alone because time (as conceived of here) has but one dimension.

Noninteracting time and space

Einstein's Special Theory of Relativity galvanized the scientific community and changed the way we think of space and time. However, cartographic time harks back to the simpler Newtonian conception of noninteracting space and time. While it may well be that a geographer's spatial and temporal dimensions do interact (see, for example, Isard, 1971), spatiotemporal data are more generally useful when space and time are recorded separately. Individual researchers can design models or representations of hypothesized space-time interactions to operate upon the absolute temporal and spatial coordinates stored in a cartographic representation.

Temporal boundaries

For good or ill, cartography traditionally describes space by the boundaries between its entities or attributes. While boundaries severely simplify and perhaps misrepresent phenomena in the physical and social world, they are also a simple and compact manner of expression. Taking a cue from these spatial practices, we can also describe time simply and compactly by the boundaries between its entities or attributes.

Spatial boundaries form when adjacent locations differ; temporal boundaries form when adjacent states of the modelled system differ, i.e. when change occurs. It is reasonable, then, to consider change to be the essence of cartographic time. This view of time as a series of abstracted events is not unlike that of Aristotle, Liebnitz, and the majority of information-science researchers.

Convention allows us to draw spatial boundaries firmly, regardless of

---

[5]Spatial stereotypes of time are quite common. Aaronson (1972) questioned 226 college students on the representational structure of space and time. The students overwhelmingly placed the past to the bottom, left, and back, and the future to the top, right, and front, with the present running between the two.

whether they represent gradual transitions or formal lines of demarcation. While temporal boundaries are no more discrete than are spatial ones, sharp lines also prove useful in their representation (Figure 3.1). Just as choropleth classes help simplify attribute distributions, so temporal episodes can simplify time.

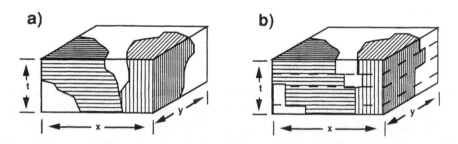

Figure 3.1 Discrete boundaries between time slices stored in a cartographic representation.

The language of temporal maps

Among the thornier issues of temporality is defining and describing time and its passage. Mandelbaum (1984) eloquently describes the difficulty of distilling a universal 'now' from the 'evanescent present' and argues for defining an artificial present as a reference point from which the past and future depart. A temporal GIS has a particularly pressing need for an artificial present because a system can establish a 'base state' to regard as the present, regardless of whether that nominal present coincides with the real-world present. Because of temporal ambiguities, I have chosen the linguistic concept of tense—present tense, past tense, and future tense—to describe data's relative location along the time line. These terms sidestep the implication that a base state is always the most current data or the 'evanescent present'. Rather, tense implies a viewpoint—in this case, the viewpoint of the system.

Many have likened the map to a linguistic system. The usefulness of linguistic terminology in describing cartographic concepts has been explored in depth by Dacey (1970), Young-nann (1978), Nyerges (1980b), Gerber (1981), Child (1984), and Head (1984) and has been touched upon by Robinson and Petchenik (1976), Arnberger (1974), and Morrison (1974). Head argues that cartography has more in common with natural language than with the communication models based on electronic information theory, although Robinson and Petchenik (1976) disagree, since maps lack an analog for either words or syntax. Child's (1984) comparison of maps to poems is perhaps more reasonable; the meanings of map symbols tend to be allegorical, as word meanings are in poetry.

Of interest to cartographic time are the 'verbs' of the language that express not just what occurs, but also the relationship of that occurrence to now, i.e. the verb tense. 'To be' is the most prominent map verb (i.e. 'the forest *is* beside the road' or 'the largest lakes *are* north of town'), although 'may be' sometimes softens the certainty of 'is'. Both the Defense Mapping Agency and the US Geological Survey employ special symbols to describe specific instances of uncertainty, such as form lines in lieu of contours, indefinite or unsurveyed shoreline symbols, and annotations (e.g. 'relief data incomplete' or 'limits of reliable relief information'). In some cases, areas of the map are left entirely blank. Other methods of expressing uncertainty include notes in the margin (e.g. 'this information has not been field checked') and reliability diagrams.

The vast majority of map information is expressed in the present tense. However, exceptions exist. Elderly publication dates on contemporary maps can imply a voice from the past speaking in the present tense; antique or historical maps speak of geography and cartographic methods in the past tense. Maps depicting temporal themes employ a series of tenses. Map symbols or annotations that describe planned or anticipated changes supply a rudimentary future tense, and the measures that imply uncertainty or imprecision supply a conditional tense. To actively apply the construct of verb tense to maps is tantamount to adding a 'when' component to the 'what' and 'where' components described by Head (1984). In Sinton's terms, 'tense' permits us to measure time, albeit ordinally.

In addition to existence, maps describe entities. Each entity within the system modelled by a map or GIS has a lifespan, as does the system itself. To enliven the discussion, I personify these entities by delimiting their lifespans with birth, death, and (when applicable) reincarnation. These terms serve to distinguish changes in the modelled system from assumed actions by the computer; 'addition' and 'deletion' are, in this way, reserved to describe data-processing procedures. For the same reason, Chapter 6 discusses methods of 'retiring' rather than archiving data, since 'archiving' bears specific data-processing connotations that it would be premature to adopt.

Temporal objects

Cartographic time is populated by objects that also require definition and terminology. Drawing comparisons between spatial and temporal entities is one useful descriptive tool. Table 3.1 summarizes some analogies between space and time, building on those offered by Parkes and Thrift (1980).

The temporal parallel of 'map' is 'state'. This term for aggregate conditions arises from general systems theory and its roots in mathematical physics, which considers the history of a system to be a series of states

Table 3.1 Parallels in spatial and temporal constructs.

	Cartographic space	Cartographic time
Overall configuration	map	state
Configurations separated by . . .	sheet lines	events
Regular sampling units . . .	cells	hours, days, decades, etc.
Meaningful units	objects	versions
Separators between units	boundaries	mutations
Size measured by . . .	length, area	duration
Position described by . . .	coordinates	date
Contiguous neighbours . . .	adjacent objects	previous and next versions
Maximum number of neighbours	infinite	two

punctuated by 'events' that transform one state into the next. A cartographic state consists of a spatial configuration of objects, each of which can change somewhat independently of the others. Just as a map is transformed from state to state by events, an object is transformed from one 'version' to the next by 'mutations'. Thus, each map state freezes geographic evolution into a configuration of object versions.

The topology of cartographic time

The view of temporality shared by most researchers envisions a sequence of states punctuated by events that transform one state into the next. States have duration, and are therefore represented by time intervals. Some conceive of events as instantaneous and thus represent them as points; others consider them to be truncated but potentially telescoping intervals (for example, Allen, 1983; 1984). Within the simple framework of discrete but linear time, a number of issues arise.

Each object mutation is an event that produces a new object version and map state (Figure 3.2). This interrelationship results in a topology comprised of many parallel lines—a view of time that coincides with that of other time specialists. For example:

> Given that temporal information is expressed in space, and that intervals between events are subject to variability in the recording . . . we can begin to envisage a topology of temporal information in which we have rubber strings or nets or sheets on which events are recorded as knots or other singularities, expressing certain necessary relations between events,

relations which hold good even though the rhythms of all the clocks concerned are variable (Meredith, 1972, p. 260).

Aside from the obvious interrelationships between object versions and map states, the succession of each object's versions and mutations (or one map's states and events) has an internal structure. A version or state can be seen as a line segment that represents the duration of a condition, while a mutation or event is a point that terminates that condition and begins the next, forming a zero-dimensional boundary between two one-dimensional 'regions'. Thus, temporal units that share a boundary can be considered contiguous neighbours in time (Figure 3.3).

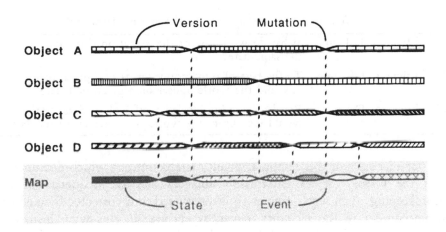

Figure 3.2 The topology of cartographic time. Relationships of objects to maps, mutations to events, and versions to states.

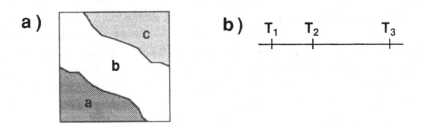

Figure 3.3 Spatial and temporal neighbours. (a) The spatial neighbours of b are a and c. (b) The temporal neighbours of T_2 are T_1 and T_3.

Three facets of cartographic time

An important distinction for information processing is the difference between when events occur in the world and when the database records them. These two facets of time are called 'valid' and 'transactional' by Snodgrass and Ahn (1985), 'logical' and 'physical' by Lum *et al.* (1984), 'extrinsic' and 'recording history' by Ariav (1986), and 'object level' and 'system level' by Bolour and Dekeyser (1983). I use the terms 'world' and 'database' time, which I coined because they are simple and independent of specific data-processing connotations. Database time traces the history of database transactions and world time traces real-world events. Table 3.2 clarifies the distinction.

Table 3.2 Information provided by tracing both world and database time.

World time = DB time	What do we know of the world on this date, as of the same date?
World time < DB time	To the best of our knowledge as of a given date, how did the world appear on a given past date?
World time > DB time	Based on our knowledge as of a given date, how would we expect the world to appear at a later date?

A familiar user of this temporal duality is accounting, which documents both the history of an enterprise's finances and the sequence of its recordkeeping. If an accounting error is discovered, it is corrected by posting an amendment to the accounts, never by erasure. In this way, historical information can be altered if knowledge of the past improves, but an ineradicable paper trail tracks the change. Unless a database traces both world and database time, it cannot describe events that did not occur as scheduled or current events that affect the past. Inclusion of both world and database time is particularly important to a decision-support system. For example, unless a database can be viewed in its state when a decision was made, that decision may appear to be flawed if data were altered postactively.

Cartography can be divided into three levels of abstraction: entities, objects, and symbols (Nyerges, 1980a; 1989). Change at the entity and object level is clocked in world and database time, respectively. However, change at the symbol level is clocked in display time (i.e. through animation, symbolization, or update). Display time, being independent of database representation, is discussed no further in this work.

Constant identity

Because temporality involves change, a fundamental problem is how to identify all states of a changing entity as versions of the same entity. The mechanical problem of relating entity versions in a database is easily solved by using surrogate values rather than natural keys to serve as unique identifiers. But the logical and conceptual problems remain: what makes an entity a version of another rather than a new entity altogether? For example, after painting a lighthouse most would agree that it remains the same entity. Is it the same entity, however, if the luminance or elevation of the light is changed? If the building is moved five metres back from the water or five miles down the coast? Or if it is torn down entirely and rebuilt in the same location? Clearly, any or all of an entity's attributes can change over time, with some associated granularity of change. Additionally, each object has an associated lifespan in which it is born, dies, and (possibly) is reincarnated. For this reason, it is crucial to define, for a given application, the essential elements that identify each of its entities.

Incremental attribute change

In addition to having finite lifespans, entities in the real world can change in attribute, shape, or location. The next chapter discusses shape and location change because their treatment is closely tied to a representation method and the type of data being treated. Attribute change alone is universal.

Change can involve one or more attributes. Some attributes never change, some attributes change over time, and yet others are measured in the time domain (Clifford and Tansel, 1985). Note that an attribute's category depends on what it describes. The attribute 'address' may be constant to a house but time-varying to a person. And the attribute 'time' varies with time and is measured in the time domain when it describes the moment a flight leaves the gate each day.

Attributes measured in the time domain can be uninterpreted values, e.g. the scheduled arrival time of a flight, or they can be time stamps that clock entity changes. Segev and Shoshani (1987) enumerate three possible relationships of time stamps to objects and their attributes. In brief, an object may have only one attribute set per time slice; a set of objects may be defined always to share an attribute set, even when the attributes change; or an object and its attributes may be clocked by several measures, for example when a system traces both world and database time.

Scale independence of spatiotemporality

Cartographic time encapsulates the temporality of the physical world. For this reason, cartographic objects change at different rates because phenomena change at different rates. The ideal temporal scale for viewing phenomena depends on the phenomena; ultimately we need methods of deriving one temporal scale from another.

Common wisdom holds that information should be captured and stored at the finest resolution required. By this approach, the system supplies algorithms to 'generalize' or 'coarsen' the data when an application needs a smaller-scale representation than is stored. The desirability of storing spatial information in this manner is a recurrent theme in geographic and cartographic works, as witnessed by energetic research that spans over twenty years (Perkal, 1956; Tobler, 1964; 1970; Topfer and Pillewizer, 1966; Srnka, 1970; Sukhov, 1970; Rhind, 1973; Poiker, 1976; Christ, 1978; Buttenfield, 1986; Beard, 1987; and Nyerges, 1989, are but a sampling).

Anticipating what will be the finest required resolution is, of course, critical, and equally so in the temporal domain. Isard (1970, p. 19) comments that the 'choice of a relevant transformation of time resembles the choice of relevant map scale'. A poor choice of scale could give the false impression that two objects coexisted at the same time, when in fact they never did.

Generalizing or interpolating over time also is a challenge, and could easily provoke a level of research excitement comparable to that generated by spatial generalization and interpolation. While temporal scale change is likely to be simpler than spatial scale change because of time's single dimension, some commonality between the two may exist.

Temporal data are comprised of only points (events) and the lines connecting them (states). However, the points stored by the system can be considered either discrete events (e.g. a shoreline slumping overnight into a bay) or samples of a continuous process (e.g. a shoreline gradually eroding over time, but surveyed only periodically). The former temporal data type is most aptly generalized using a categorical point-selection approach (e.g. Kadmon, 1972; Stenhouse, 1979; Langran and Poiker, 1986). Conversely, a continuous process would be more suited to a line generalization approach that selects values that deviate from a norm by some established tolerance (e.g. Douglas and Poiker, 1973).

It does appear that temporal and spatial scale are loosely associated; several theories to this effect have been advanced. Isard (1970) equates time with distance, and suggests that the two be scaled in tandem. Parkes and Thrift (1980) extend this construct by pairing temporal and spatial scales into distances that could be traversed in time spans (e.g. minute/room, day/city, week/region, year/nation, and life/world). Also Whitelaw (1972) scales times and spaces (in a method similar to that of Parkes and Thrift) to activities.

A capability to produce many scales from a single-scaled database could facilitate research on how micro and macro phenomena relate. Many have noted that both regional and local components of space-time models exist (e.g. Olsson, 1968; Haggett, 1971; also see Burrough, 1987), with problematic relationships between scales and different wavelengths at different scales. Because of this, Morrill (1982) suggests that we study large and small areas intensively, and also long and short periods intensively. Indeed, the mechanisms that relate phenomena at different scales are poorly understood, as evidenced by the findings of Chaos science (Gleick, 1987). Fractal behavior of both spatial and temporal phenomena has been observed (Mandelbrot, 1977; Goodchild, 1980; Burrough, 1981; Armstrong, 1986; Mark and Aronson, 1984). Proper design of a temporal GIS and faithful capture of spatiotemporal information could one day produce a vehicle for identifying fractal behavior in spatiotemporal phenomena.

Implementing cartographic time

At the root of any digital system is a conceptual model that encapsulates the actual system for the computer. In the case of a temporal GIS, the concern is how to convey the essentials of space and time in such a way that the functions of a given application can be performed effectively. Cartographic time defines these essentials. All that remains is to devise a construct that describes these components to the computer.

Conceptions of time

A useful starting point for a conceptual model of spatiotemporality is a set of popular and recurring images of space and time. A few well-honed conceptions of time dominate the popular and research literature on the subject. Each conception offers a different view of temporality and emphasizes a different set of components. The goal here is to identify a conception that treats the components of cartographic time most effectively, and to develop that view of time as a conceptual model for a temporal GIS.

The space-time cube

Among the more evocative conceptions of spatiotemporality is a three-dimensional cube that represents one time and two space dimensions. This is the space-time model of Hagerstrand (1970), Rucker (1977), Szego (1987), and many others. Space-time cubes depict processes of two-dimensional space that are played out along a third temporal dimension. The trajectory of a two-dimensional object through time creates a worm-like pattern in this phase space.

It is conceivable that methods could be developed so that a temporal GIS could treat spatiotemporal objects as hypothetical solids within a space-time cube. To access information from this cube would entail referencing a point, tracing a vector, slicing a cross-section, or trimming a smaller cube from within the universe cube. Each of these operations would be progressively more complex, and the difficulty would increase with data volume.

CAD systems model three-dimension solids as described above, but this technology was not designed to cope with the thousands or millions of topologically connected objects of geographic space and time. The fact that this approach would tax hardware and would require entirely new spatial logic and digital algorithms places it beyond practical reach for a near-term temporal GIS capability.

A three-dimensional approach also would entail a great deal of philosophical and conceptual soul-searching on the part of cartographers. For example, how would this construct describe the temporal topology of states, events, objects, and mutations? Devising a method to encode fully three-dimensional topology is not a trivial pursuit. And how would we represent the third (temporal) face of a changing two-dimensional object? Do we assume that change is gradual and evenly paced, and interpolate a sloping face between one observation and the next (as in Figure 3.1a, above)? Or do we adhere honestly to what we know and describe the changes in a stepwise fashion to reflect their measurement (as in Figure 3.1b, above)? Many problems of this sort would face a three-dimensional GIS development. If users are to enjoy a working temporal GIS in the next several years, GIS developers must exploit proven frameworks and not overstep expected technical gains.

Sequent snapshots

A second common model of spatiotemporality is a snapshot sequence of time slices (Figure 3.4). The nature of each time slice is captured by Wood and Fels (1986), who describe maps as sequent snapshots that record only fixed phenomena because moving phenomena become transparent blurs on the film. Based on the preceding discussions, we can consider that such a snapshot would capture all phenomena whose granularity of change approximates its exposure time.

Time-slice snapshots are an intuitively appealing space-time model. They have roots in traditional mapping and mimic the progressive nature of a slow-motion video. They are in current use to approximate spatiotemporality in a GIS (e.g. Dangermond, 1984; Ross, 1985). However, snapshots are a crude means of representing the most important component of cartographic time: change. Each snapshot describes what exists at T_i. But

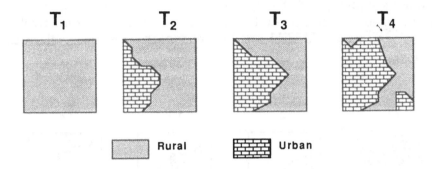

Figure 3.4 Time-slice snapshots representing urban expansion into a rural area. Intervals between time slices are not necessarily equal.

to detect how T_i differs from T_j, the two snapshots must be compared exhaustively.

The root of the problem is that snapshots represent states but do not represent the events that change one state to the next. In many respects, time-slice snapshots are the temporal equivalent of the formless spaghetti data structure. In both cases, the database objects reflect the graphic, not its underlying meaning. By storing only absolute locations (unaccompanied by relative position), the questions 'where is it?' and 'what is it?' are easily answered but 'where and what are its neighbours?' (whose temporal translation is 'when do the previous and next changes occur?') are not. Interestingly enough, the three major disadvantages of the spaghetti method of recording spatial information—hidden structure, no error trapping, and redundant storage—match the shortcomings of the snapshot method of recording temporal information.

- Hidden structure. An object's temporal neighbours are its previous and next manifestations, located at the moments where the changes occur. Because snapshots capture states, not versions, the boundaries between versions are difficult to locate.
- No error trapping. With no understanding of or constraints upon temporal structure, it is difficult to devise or enforce rules for internal logic or integrity.
- Redundant storage. Regardless of the magnitude of change, a complete snapshot is produced at each time slice, which duplicates all the unchanged data.

Base state with amendments

A third common image of geographic time is that of a base state with amendments superimposed (Figure 3.5). Because this construct provides a terse description of change—a fundamental component of cartographic time—refinements should yield a viable conceptual model upon which to build a temporal GIS.

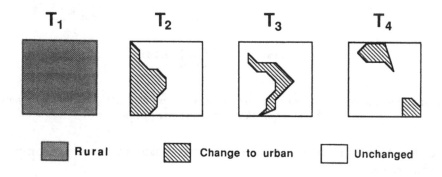

Figure 3.5 Base state with amendments that describes the urban encroachment of Figure 3.4.

The importance of using storage economically should not be underplayed. Temporal data, with its added dimension, will always be bulkier than its atemporal counterpart. Because spatial data alone tax computer system capacities, spatiotemporal data appear to be completely overwhelming. Volume can be reduced considerably if temporal data are stored, not as complete new snapshots of the study area, but as departures from a base state. The type, timing, and sequence of change are the essence of temporality. For this reason, storing change data rather than sequent snapshots also makes far more sense logically.

Change-only, or event-based, temporality springs naturally from the object-based philosophy of the vector GIS school (see Peuquet, 1988). To wit, objects or events are stored as they occur rather than at regular sampling intervals. For example, an object-based space might be described succinctly as 'a lake lying within a forest'. An event-based temporality might be described as 'the forest becomes a meadow at time t'. In contrast, a sample-based description of the same space and temporality might be 'a forest pixel, another forest pixel, a water pixel, another water pixel, another water pixel . . .' and (describing a single pixel sampled over time) 'a forest pixel, still a forest pixel, still a forest pixel . . .'.

Because it represents change as the boundaries of both states and versions, the base state with amendments is superior to snapshots. Just as the astructural style of snapshots shares the spaghetti structure's problems, the temporal structure of this construct mimics the topological structure's assets.

- Temporal structure is evident. Temporal neighbors (i.e. a version's previous or next forms) are located by finding the mutation that separates them.
- Errors can be trapped because improbable events can be identified and the geometric information accounts for all space, which enforces internal consistency.
- Redundancy is minimal because an object version is stored only once.

Space-time composite

In essence, the base state with amendments is constructed by flattening the three-dimensional space-time cube into two spatial dimensions. Differences in the time dimension show up as new objects in two-dimensional space. For storage purposes, the base state with amendments serves as a starting point to build a space-time composite of accumulated geometric changes.

Figure 3.6 shows a space-time composite of Figure 3.4's snapshots. Each change causes the changed portion of the coverage to break from its parent object to become a discrete object with its own distinct history. In other words, the representation decomposes over time into increasingly smaller fragments—the area's greatest common spatiotemporal units—each of which references a distinct temporal attribute set. This method of temporal decomposition was originally suggested by Chrisman (1983) and is described in detail in Langran and Chrisman (1988).

By identifying units with coherent histories, the mutation of those units can be described by aspatial attributes. In sum, generating a space-time composite of a time sequence's geometry and topology reduces the three dimensions to two and permits us to treat space atemporally and time aspatially.

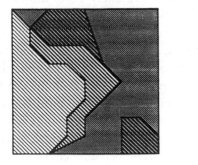 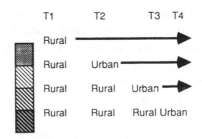

Figure 3.6 A space-time composite of urban encroachment. Each polygon has an attribute history distinct from that of its neighbours.

The mechanics of space-time compositing begin with a base map that represents a region's geometry and spatial topology at some starting time. Each database update session generates an overlay, as described previously for the base map with disjointed overlays. Once accepted for permanent inclusion (having passed error-detection procedures), the overlay is incorporated into the system using the same intersection procedure currently used for polygon overlay (Dougenik, 1980). New nodes and chains are added to the historical accumulation, forming new polygons that have attribute histories distinct from those of their neighbours. Each unit's attribute history is represented by an ordered list of records. A record contains an attribute set and the time when that attribute set is valid.

Accessing temporal information stored in the space-time composite is conceptually straightforward. To compile a single time slice from the composite, one has only to 'walk' the history list of each polygon to locate the attribute set that was current at the desired time slice. If polygon neighbors in the time slice share a single attribute, the chain that separates them is dropped.

Composited versus uncomposited space-time

Among the assets of the topological model is that it accounts for all space within the database universe and thus facilitates error detection in spatial databases. Spatiotemporal error detection is even more challenging because error potentially propagates over time. The space-time composite, which employs the topological model, similarly accounts for all space and all times by requiring all objects to have an attribute set for any given moment.

However, the space-time composite approach does suffer two unfortunate side-effects: the representation decomposes into progressively smaller objects, and the identifiers of changed objects must be altered retroactively.

Decomposition. Unchecked decomposition ultimately poses intolerable storage problems for any system. Chapter 6 discusses options for managing the volume of spatiotemporal data. The alternative to decomposition, i.e. maintaining unintersected or outdated objects, is far worse because it creates disunity across time slices and violates the space-exhausting principle that has proven effective in recognizing error. Uncomposited approaches are also far more complex. For example, we could permit objects to be born and die when they actually appear in the representation. This would require nodes, chains, and polygons to have effective dates, chains to have variable-length lists of polygon left/right, and polygons to have variable-length lists of temporal rings, as follows:

Temporal Node: ID, birth, death

Temporal Chain: ID, birth, death, node from, node to, polygon left
                   list, polygon right list
Temporal Polygon: ID, birth, death, temporal ring list

The polygon lists and temporal ring list both include effective dates of each element.

Uncomposited approaches also would demand complex data structures to trace the ancestry of objects, since a new chain could break an existing chain or polygon into two. Would the two children of a broken object bear new identifiers that would be cross-referenced to the parent, or would one be considered the same object and the other considered new? Either approach is unwieldy and not recommended.

Retroactive changes to identifiers. Each time the space-time composite splits an object into two, the old object is effectively replaced by two new objects with new identifiers.This means that throughout the database, each occurrence of the old object identifier must be replaced with one or both of the new ones.

The prospect of retroactively changing identifiers throughout the database is troubling. An identifier can be referenced by many objects or features, but no practical alternative exists. To ameliorate the effects of retroactive changes to identifiers, changes can be limited to partitions (as described in the next chapter) and updates can be speeded by navigation, rapid-access methods, and dataflow algorithms.

The space-time composite and cartographic time

Table 3.3 summarizes the components of cartographic time that this chapter introduced. Each can be treated within the space-time composite framework. More ambitious definitions of cartographic time are likely to require alternate or extended frameworks.

Table 3.3 The components of cartographic time.

Time is the fourth cartographic dimension.
Cartographic time and cartographic space do not interact.
Temporal boundaries are sharply drawn.
The language of maps extends to temporal constructs.
Spatial objects have temporal corollaries.
Cartographic time's topology is that of many parallel lines, one per object.
Cartographic time has three facets: world time, database time, and display time.
Objects maintain their identity despite change over time.
Objects can change in different ways.
Temporal objects can be treated irrespective of scale.

The space-time composite treats the fourth cartographic dimension as it does the third: in separate attribute structures. Segregated in two databases, cartographic space and time do not interact in the space-time composite. Time stamps comprise discrete temporal boundaries that mark the event that separates two object versions. Surrogate values uniquely identify entities as they change over time. Data of any scale can be treated within one space-time composite. Chapter 5, which discusses the temporal (attribute) database, describes how to treat world and database time.

4

Implementing the approach on common data types

Chapter 3's presentation of the conceptual model revolved around polygon coverages. But ideally, a single conceptual model should be able to adapt to all geographic data types. This chapter uses four common data types as case studies to extend the space-time composite to specific representational needs.

Continuous surfaces

Two major data structures are used to describe surfaces. The simplest and most common is the regular grid. The triangulated irregular network, an alternate structure for describing surfaces, is not treated here.

Grids describe a surface via samples taken at regular intervals, forming a rectangular matrix. Elevations are derived thus from photocompilers; earth and map images are grids scanned by imaging systems. Gridded data of finer resolution are often referred to as 'raster' or 'image' data. For simplicity, I refer to all data sampled at regular rectangular intervals as gridded.

Gridded data are location-based. They describe what exists at a given location (to a given resolution) but do not easily yield such information as perimeter, shape, or contiguity. However, because this information is not always of overriding interest, and because they are easily collected and manipulated, grids are a prominent and popular geographic data type.

Not all gridded data are stored sequentially, although they are generally processed as such. Most gridded data are stored in row order (Figure 4.1a). This is particularly true of raster data because the specialized array processing hardware used in image display and analysis is geared to row-ordered data. But row ordering is not ideal for all purposes. Some applications are better served when measurements that are collocated in the living world are also collocated in storage. A small step in this direction is row-prime ordering (Figure 4.1b). Row-prime ordering does minimize the storage gap between row elements at matrix edges, but no coherence is gained between column

45

elements. Hierarchical ordering is one proposed solution to the collocation problem (Figure 4.1c), although Goodchild and Grandfield (1983) demonstrate that neither row-prime nor hierarchical orderings improve on the spatial autocorrelation of data in storage provided by the simpler row order.

Storage structures for row order and row-prime order are sequential. Some applications use a hashed indexing scheme (e.g. Morton, 1966) to improve access to the desired row or column. Hierarchical grids are generally stored in quadtrees, which are organized using pointers or computed addresses (line quadtrees).

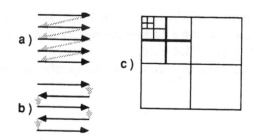

Figure 4.1 Ordering gridded data. (a) row order, (b) row-prime order, (c) hierarchically ordered.

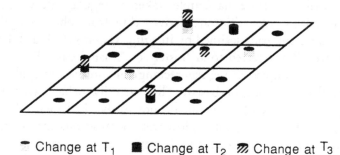

‸ Change at T_1 ■ Change at T_2 ⧄ Change at T_3

Figure 4.2 A temporal grid represented as a space-time composite. No new spatial objects are formed because no intersections occur. The result is a set of variable-length lists referenced to grid cells.

According to Sinton's (1978) scheme, a grid controls location and measures theme. For this reason, a grid of given resolution and extent can undergo attribute change only. In that case, generating a space-time composite from a sequence of grids is straightforward because the grid cells (theoretically) align and produce no geometric intersections.[6] In essence, the values held by each grid cell over time are described by a list of attributes bracketed by validity dates (Figure 4.2). If grid characteristics are redefined to sample at a different resolution or cover a slightly different area, a space-time composite would cause new cells to be born from the intersection of the two grids.

Categorical coverages

Another common geographic data type is the exhaustive polygon coverage. Polygon coverages offer location-based information: all space has an assigned value. But polygon coverages divide space into cohesive zones, rather than arbitrary samples, thus forming an intriguing mix of location- and feature-based approaches.

Changes to polygon coverages are attribute changes that manifest themselves in a new set of polygons. We can conceptualize a temporal polygon coverage as a set of polygons, each with an attribute history distinct from its neighbours. The concept of the space-time composite was introduced and demonstrated using a polygon coverage. Figure 4.3 shows snapshots of a polygon coverage and its accumulating objects in the space-time composite.

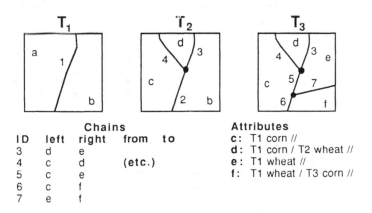

		Chains			Attributes
ID	left	right	from	to	c: T1 corn //
3	d	e			d: T1 corn / T2 wheat //
4	c	d	(etc.)		e: T1 wheat //
5	c	e			f: T1 wheat / T3 corn //
6	c	f			
7	e	f			

Figure 4.3 Temporal land cover data represented as a space-time composite. T_1, T_2, and T_3 represent the database state at those times. Superseded database states are recoverable by recomposing objects but are not retained in their original form.

---

[6]If grid cells do not align, they must be resampled at high cost to spatial and attribute accuracy.

Networks

Networks describe the links between nodes in the physical world. Unlike Corbett's (1979) topological model (which treated graphs), the edges of the real-world networks need not intersect at nodes.

Creating a space-time composite to represent a network is straightforward. The treatment described above for categorical coverages applies to networks except the attributes reference chains and nodes, not polygons. Figure 4.4 shows a space-time composite of a temporal stream network.

Geographic features

Geographic features are the stuff of which reference maps are made: houses, roads, rivers, lakes, towns, airports, and other units that hold particular

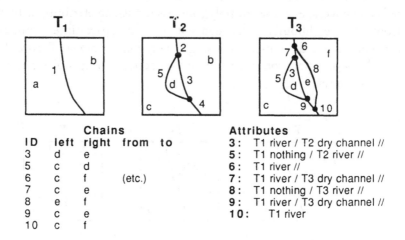

Chains

ID	left	right	from	to
3	d	e		
5	c	d		
6	c	f	(etc.)	
7	c	e		
8	e	f		
9	c	e		
10	c	f		

Attributes

3: T1 river / T2 dry channel //
5: T1 nothing / T2 river //
6: T1 river //
7: T1 river / T3 dry channel //
8: T1 nothing / T3 river //
9: T1 river / T3 dry channel //
10: T1 river

Figure 4.4 A temporal stream network represented as a space-time composite. The states at T_1 and T_2 can be recovered from the database version stored at T_3 but their original identifiers are not retained.

meaning to a given application. Dacey (1970, p. 79) recognized the peculiar nature of features when he wondered 'how to specify identity conditions within a single semantic system for mobile objects whose non-positional attributes are invariant over time but occupy a temporal series of spatial regions' (i.e. features) 'and geographic regions whose nonpositional attributes change over time' (i.e. objects). In essence, Dacey defines features according to the Sinton framework: the identity (attribute) is fixed.

Lehan (1986, p. 25) defines a feature as a 'physical entity that is recognized in the user's definition of reality' and reality as 'the characteristics of a region that are significant to the user of a specific classification of spatial information'. By this definition, 'features' facilitate communication between a system and its users; they index entities with particular meaning to the application to build a pathway directly to that entity. The Technology Exchange Working Group (Guptill, 1988) defines 'feature' to be 'a set of phenomena with common attributes and relationships'. In contrast, the proposed US National Standard (Morrison, 1988) includes the digital representation as part of a feature.[7]

What finally distinguishes features from objects? Both are described by points, lines, and areas, and no single representational model is prescribed. The answer lies in the data's character and application. The character of grids and polygon coverages is locational—they describe the attributes of two-dimensional space. A grid controls location by sampling space at regular intervals and a polygon coverage controls attributes by categorizing their possible values. Nonetheless, in both cases the data report on locations.

Conversely, features are thematic units, often named, that are comprised of objects. Feature data describe the attributes of geographic entities, which in turn occupy locations in space. Unlike objects, features do not fill a region exhaustively. Features can be composed of discontinuous points, lines, and areas and can be represented by many nodes, chains, or polygons, possibly noncontiguous. For example, the feature 'Washington State' is comprised of its mainland, the San Juan Islands, other assorted islands, and Point Roberts (a noncontiguous peninsula). Glacier Park is comprised of its area, its boundary, the Going-to-the-Sun Road, lakes, streams, ranger stations, and various tourist facilities. The national park system is comprised of many nonadjacent polygons located throughout the US.

The application of data further distinguishes features from objects. Consider how a geologist and a mining company would view a geological cross-section. The geologist who sought a gestalt understanding of the rocks and minerals within the volume would treat the geological data as objects with attributes. Conversely, the mining company would be more likely to focus on ore deposits or veins, which would be treated as features embedded within the spatial objects and their attributes. In summary, what makes a feature a feature is a system's users and usages.

---

[7]The National Standard defines 'feature' to be 'a defined entity and its object representation', defines 'object' to be 'a digital representation of all or part of an entity', and defines 'entity' to be 'a real-world phenomenon that is not subdivided into phenomena of the same kind.'

White's (1983) distinction between geography and geometry is pivotal to this discussion. According to White, geographic information could be a boundary trace of a political jurisdiction, while geometric information could be the links and nodes of a network that separates an entire region's political jurisdictions. White's geographic data are feature-based because entities are singled out to be of particular interest. White's geometric data are location-based because the configuration and coverage of information is of primary concern. Feature-based data are stored at a thematic feature level, while location-based data are stored at an object level, to which attributes are keyed. I use the term 'level' purposely; the object and feature levels are closely tied. One can obtain object-level information through the feature level, and vice versa. These two levels merely provide different entry points and complementary views of the data.

Features can be represented without employing exotic processing procedures. For example, a location-based approach could treat the feature 'national parks' by creating a 'coverage'[8] with the categorical attribute 'national parks'. At the simplest level, each polygon in this coverage would be assigned the value 'national park' or 'not national park'. Conversely, a feature-based approach would create a feature data structure to reference the space-filling object level.[9] This strategy avoids the overhead of maintaining symmetric attributes (e.g. balancing 'national parks' with 'not national parks'), although the feature data structure also involves overhead.

The temporal topology of objects versus features

Differences between objects and features become more pronounced in a temporal GIS. Over time, features can move, change shape, and be born, die, and be reincarnated—all of which are impossible for locations. These mutations have obvious impacts on representation, since in the space-time composite, geometric change is represented by adding more objects to the existing ones. However, objects can undergo attribute change, which provides a basis for representing feature change as a superset of object change and accumulation.

Figures 4.5 and 4.6 clarify the distinction between the temporality of features and objects. Object temporality can be construed as a chain whose beginning and ending nodes correspond to the beginning and ending of the temporal partition the object occupies. The chain's vertices represent the

---

[8] A coverage is a digital overlay of attributed spatial data.
[9] The USGS's Digital Line Graphs and the Defense Mapping Agency's Standard Linear Format both superimpose a feature level upon the topological network structure using a strategy similar to that described here.

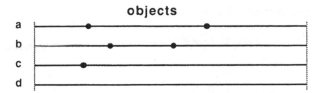

Figure 4.5 Object lifespans. The temporality of objects can be construed as a chain whose vertices represent moments of attribute change. The endpoints of an object's temporal chain coincide with the boundaries of its temporal partition.

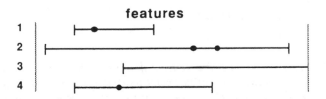

Figure 4.6 Feature lifespans. The temporality of features can be conceived of as a chain whose vertices represent the moments when the feature changed. The endpoints of a feature's temporal chain correspond to the feature's birth and death, and are independent of the partitioning. Vertices represent shape change, movement, and attribute change.

moments when the object undergoes attribute change (Figure 4.5). Objects cover their spatial and temporal partitions exhaustively.

Exhaustive coverage is a familiar concept, since that is a current method of detecting spurious spatial data. The space-time composite extends exhaustive coverage to the temporal domain. When a geometric change to the area causes an object to break into new greatest common units, the representation behaves as if the new objects existed with null attributes since the beginning of the temporal partition. Figures 4.2 and 4.3 illustrated object temporality in the space-time composite framework for a categorical coverage and a stream network.

Unlike objects, features do not necessarily fill their spatial partition exhaustively. Similarly, the chain that describes a feature's temporality does not necessarily begin and end at the boundaries of a temporal partition (Figure 4.6). Features can undergo shape change, movement, birth, death, and reincarnation.

A representation strategy for temporal features

If feature-based and location-based information are complementary, as I

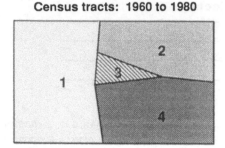

Census tracts: 1960 to 1980

	population		
	1960	1970	1980
A			
B			
C			

1   1960:Tract A //
2   1960:Tract B //
3   1960:Tract B / 1980:Tract C //
4   1960:Tract C //

Figure 4.7 Temporal census-tract data represented as a space-time composite. Polygons 1, 2, 3, and 4 are the greatest common spatiotemporal units. Polygon 3 was part of Tract B until 1980, when it moved to Tract C. Census statistics such as population require no special treatment because time is symmetric with respect to other attributes.

argue here, then an integrated means of dealing with the two is essential. One strategy is to maintain an exhaustive object-level coverage to represent location-based information and reference the objects in that coverage from a feature-level superstructure.

Figure 4.7 shows a space-time composite of temporal census-tract data. Polygons 1, 2, 3, and 4 are the greatest common spatiotemporal units of Census Tracts A, B. and C. Until 1980, the three census tracts were represented by three polygons because Polygon 3, being part of Tract B, was incorporated into Polygon 2. Then in 1980, the area represented by Polygon 3 moved from Tract B to Tract C, creating a new polygon with a distinct history.

A location-based query to these data would ask, 'What is the state of Area One at Time T?' We can reconstruct a location-based time slice from the space-time composite by collecting the objects falling within the requested area, then referencing their attribute histories to find the attribute sets that were current on the requested date. Conversely, a feature-based query would ask, 'What is the state of Tract A at Time T?' To respond, the system locates the record that describes Tract A (a feature) at the requested time, then collects the objects and attributes that it references.

Both tacks are effective for small areas with shallow histories, but we can reasonably ask whether the data processing burden is manageable given realistic geographic data volumes. Clearly, a complex query could swamp a system without an access scheme to boost performance. Chapter 7 addresses the design of such a scheme.

In sum, feature-level representations superimpose an added level of information upon the space-time composite. This same construct can be mimicked at the object level by creating what is essentially a feature coverage. However, this entails the additional bulk of identifying both features and 'nonfeatures' to maintain symmetry and logical consistency. In essence, a feature-level representation stores a set of indices into the object-level representation to reference the identity and components of individual features. A feature-level superstructure to the stream network depicted in Figure 4.4 might appear as:

$$\begin{aligned} \text{White River:} \quad &\text{T1 } 3, 6, 7, 9, 10 \\ &\text{T2 } 5, 6, 7, 9, 10 \\ &\text{T3 } 6, 8, 10 \end{aligned}$$

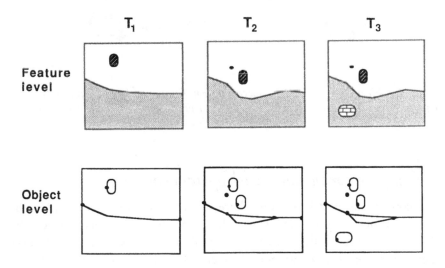

Figure 4.8 A system's view of objects and features. The object-level progression is shown in database time; 'T3' shows how the region appears to the database at the object level as of that time. In contrast, the feature-level progression shows how the world would appear to us at that time slice.

Figure 4.8 summarizes the system's view of objects and features. The feature level shows only the entities included at a given time slice in their current form. The object level shows all nodes, chains, and polygons that comprise all entities over time. Of course, at any given time slice, many objects have the attribute 'background' or 'null'.

One could question the validity of recording intersections of features that move, change shape, and are reincarnated because of the peculiar logic involved. The space-time composite can cause different versions of a feature to intersect, e.g. when it changes shape (Figure 4.9a) or moves (Figure 4.9b). Space-time intersection also can cause features from non-adjacent time slices to intersect (Figure 4.9c). As argued in Chapter 3, however, full intersection is crucial to detect errors in the data. In addition, the polygons formed by intersections measure the degree of movement and shape change.

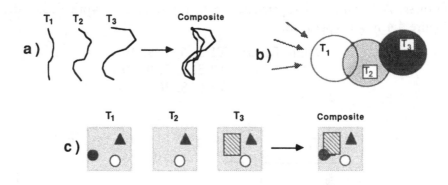

Figure 4.9 New objects created by space-time compositing of features. (a) A feature that changes shape intersects itself. (b) A moving feature intersects itself. (c) Features from nonadjacent time slices intersect.

Making the conceptual model operational

The conceptual model presented here describes spatiotemporal information and incorporates space- and time-filling concepts so errors are detectable. However, a model that works on small test datasets is not necessarily effective for realistic geographic data volumes. For this to be practical, the space-time composite requires three important components: an efficient temporal aspatial database to treat attribute information; schemes to govern clustering, volume control, and quality-control; and a method to boost the access rate of broad and temporally deep spatiotemporal data. These three topics are addressed in the chapters that follow.

5

Temporal treatment of aspatial attributes

One obvious obstacle to designing a temporal GIS is that current databases cannot trace multiple versions of entities through time. Fortunately, a great deal of recent work in computer science applies itself to this problem (Figure 5.1). Snodgrass (1986) provides an impressive survey of ongoing worldwide research activity in this area. McKenzie (1986) has produced an annotated bibliography of publications on this topic, from which Figure 5.1 was drawn. A condensed version of the discussion in this chapter is given in Langran (1989b).

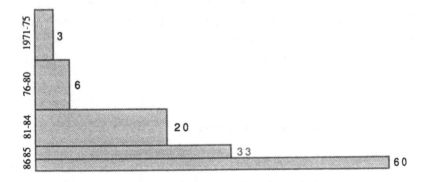

Figure 5.1 Papers on time in information processing (from McKenzie 1986)

Background

Most current research attention focuses on aspatial applications. Table 5.1 compiles the applications suggested in recent papers on the topic. But despite its aspatial focus, the work is quite germane to temporal GIS needs. Conceptual works help to build a strong framework (e.g. Sundgren, 1975; Kahn and Gorry, 1977; Sernadas, 1980; McDermott, 1982; Anderson, 1982; Schubert *et al.*, 1983; Allen, 1983; 1984; De *et al.*, 1985; Zhu *et al.*, 1987),

Table 5.1 Aspatial applications of temporal databases, as described in papers on the topic.

Maintaining medical, legal, or criminal case histories.

Recording ownership histories of automobiles, guns, etc.

Business activities such as banking, marketing, sales, plant operations, inventory
control, personnel, and reservations.

Intelligence operations requiring access to histories of political or military
activity.

Scientific experiments where measurements are collected in the time domain.

Other applications in which time is an inherent element, e.g. econometrics,
policy analysis, census and actuarial work, and music.

And most of the applied work extends the relational model in common use
by GIS.

Several factors explain the burst of interest in temporal systems. First, ten
years of widespread database use has provided users with ample opportunity
to note shortcomings. One obvious disadvantage of a digital database is that
it obscures an enterprise's evolution by overwriting out-dated information
— something that analog records do not do. Copeland (1982) notes two
other reasons for pursuing a temporal database capability. Storage costs are
plummeting, which may one day alleviate the 'agony of delete' that current
database users suffer. And, as artificial intelligence researchers attempt to
model human memory more closely, their efforts approximate a temporal
database because human memory retains past information, although its
accessibility decays over time.

Temporality in information systems

Sundgren's (1975) work in database theory incorporates time as one of the
four fundamental concepts to be modelled. Sundgren notes that the most
interesting systems are dynamic in nature. Despite this, the typical database
is a tenseless snapshot of the latest available data (Ariav, 1986). Data in the
database are temporally inconsistent because they become current at different
times. A previous database state is often retained in an archived snapshot to
restore accidentally destroyed data. This snapshot cannot support historical
or 'what if' analyses because it shares the tenselessness of the database itself.
In contrast, a temporal database models the dynamically changing world;
events are traced and no data are ever forgotten (Clifford and Warren, 1983).

Information systems can model temporality in two ways. The first,
termed 'process modelling', is the focus of this book. Process modelling
allows a system to respond to such historical or trend-analysis queries as

- Where was this feature located two years ago?
- How has this area changed over the last five years?
- Has this feature moved?
- Have these two features ever been juxtaposed?

A second type of temporal modelling, 'time modelling', is not discussed here. Time modelling is also called 'cause and effect' (Segev and Shoshani, 1987), scheduling, or 'triggering'. Management information systems use triggers to infer an appropriate action given a situation, e.g.

- if payment is not received by this date, send a second statement;
- when 25 features on the chart change, schedule a new edition.

Recent work on time modelling includes that of Barbic and Pernici (1985), Stonebraker (1986), and Struder (1986).

Temporal capabilities of current databases

Current database methods cannot express the constructs described above. Database snapshots (i.e., facsimiles) and transaction logs are only nominally temporal tools (see, for example, Abida and Lindsay, 1980; Lindsay *et al.*, 1986). Commonly, a weekly snapshot serves as a system backup and a transaction log updates the backup. But these mechanisms are not ideal for tracing database evolution. Snapshots duplicate both unchanged and changed data; of necessity, then, their temporal resolution is coarse. Snapshots also tend to be generated at regular intervals; if several events occur between snapshots, they go unnoted. In addition, temporal queries to snapshots are cumbersome because snapshots are generally stored offline and transaction logs are sequential. Finally, snapshots store states, not changes. To examine a record's changes over time, each snapshot must be referenced to see if the record has changed since the previous snapshot. In sum, snapshots and transaction logs are a clumsy way to describe database temporality. Fortunately, alternatives will soon exist.

A large body of literature now addresses temporal database design. The vast majority focuses on expanding the relational data model to incorporate some aspects of temporality. Exceptions include support of object versions in CAD (Katz *et al.*, 1986a, b); extension of hypertext to support document versioning (Delisle and Schwartz, 1986); and preliminary investigations of temporality using hierarchical (Schiel, 1983), object-oriented (Afsarmanesh *et al.*, 1985), knowledge-based (Dayal and Smith, 1986) or semantic (Urban and Delcambre, 1986) data models. However, this discussion evaluates only relational work because it has sufficient maturity for operational use.

Temporal relational database designs

A temporal database reflects changes in objects by generating a new version of the information at each object maturation. But to generate a new snapshot of the entire database for each change would create untenable storage and logic problems. The relational database provides several other options; it can generate new versions of tables, tuples, or attributes to reflect new object versions. The temporal database literature calls this procedure 'versioning'.

Conceptually speaking, we can view a temporal relational database's tables as data cubes whose depth is the time dimension. When new versions of tables are created, the cubes expand, intact. When new versions of tuples are created, the data cube has missing parts as it expands. When new versions of attribute values are created, the cube becomes quite sparse. Thus the choice of representation method is a tradeoff between storage efficiency and complexity. To create new versions of tables each time an attribute value changes creates size problems because of the redundant data. Conversely, to create new versions of attribute values alone requires multiple entries of variable number within a field, which violates the relational first-normal form and requires complex algebras to access the data.

The various approaches to temporal relational databases are showcased independently in articles and proceedings. To provide a better basis for comparison, I have implemented each approach on a common test dataset.

Table 5.2 Three snapshots of the relation HYDRO.

	Feature	X	Y	Effective	
HYDRO as of 8/30	Beacon	38	38	3/8	
	Shoal	23	32	5/2	
	Hazard	56	34	3/3	
	Bell	44	28	8/8	

9/1 transactions: The buoy is added proactively, the shoal's coordinates are altered, the light is added, and both the hazard and bell are deleted.

	Feature	X	Y	Effective	
After 9/1 transactions	Beacon	38	38	3/8	
	Buoy	45	46	9/8	(added proactively)
	Shoal	15	21	8/18	(altered) Light
		12	18	8/18	(added)

9/5 transactions: The hazard is restored and the light's coordinates are corrected.

	Feature	X	Y	Effective	
After 9/5 transactions	Beacon	38	38	3/8	
	Buoy	45	46	9/8	
	Shoal	15	21	8/18	
	Hazard	56	34	3/3	(restored)
	Light	18	18	8/18	(corrected postactively)

The paragraphs that follow discuss and illustrate major approaches using 'HYDRO', a test relation (Table 5.2). Each HYDRO record is comprised of a feature type, *x* and *y* coordinates, and clocks events in both world and database times. Changes made to HYDRO (in database time) occur on 9/1 and 9/5. The changes include current, proactive, and postactive additions, deletions, and modifications, as follows.

- The beacon never changes.
- The bell is deleted permanently.
- The hazard is deleted and later restored.
- The shoal is altered by new survey information.
- The light is altered postactively because of data entry error.
- The buoy is added proactively because of plans to place it in the harbour.

Relation-level versioning

Relation-level temporality creates and stores a new snapshot of a table when any of its attributes change. Advocates of this method have been Jones and Mason (1980); Klopprogge (1981), whose interest was in the entity-relationship model; Ben-Zvi (1982), who defined an algebra to extract a time-sliced relation from a temporal relation; McKenzie and Snodgrass (1987), who also defined a temporal relation algebra; and Clifford and Warren (1983), whose method is depicted in Table 5.3.

Table 5.3 HYDRO using the Clifford and Warren (1983) method.

DB time:	Feature	Exists?	X	Y	World time	
8/30	Beacon	1	38	38	3/8	
	Shoal	1	23	32	5/2	
	Hazard	1	56	34	3/3	
	Bell	1	44	28	8/8	
9/1	Beacon	1	38	38	3/8	
	Buoy	1	45	46	9/8	(added)
	Shoal	1	15	21	8/18	(altered)
	Hazard	0	#	#	#	(deleted)
	Light	1	12	18	8/18	(added)
	Bell	0	#	#	#	(deleted)
9/5	Beacon	1	38	38	3/8	
	Buoy	1	45	46	9/8	
	Shoal	1	15	21	8/18	
	Hazard	1	56	34	3/3	(restored)
	Light	1	18	18	8/18	(corrected)
	Bell	0	#	#	#	

Clifford and Warren's method simply time-stamps each relation. Although they do not specify whether this time stamp should be world or database time, my implementation describes world time as an attribute and time-stamps relations with database time. The main problem with this and all relation-based approaches is that replicating an entire table for each change can be quite inefficient if change is incremental. Such inefficiencies can potentially sink a spatiotemporal system that is already virtually swamped with data.

Tuple-level versioning

Several methods of tuple-level versioning appear in the literature. One by Snodgrass and Ahn (1985) employs four time stamps to bracket intervals of

Table 5.4 HYDRO using the Snodgrass and Ahn (1985) method.

Feature	X	Y	DB From	DB To	World From	World To	
HYDRO as of 8/30:							
Beacon	38	38	3/21	*	3/8	*	
Shoal	23	32	5/16	*	5/2	*	
Hazard	56	34	3/21	*	3/3	*	
Bell	44	28	8/30	*	8/8	*	

9/1 transactions: ADD Buoy and Light by appending to table. DELETE Hazard and Bell by closing their time brackets. ALTER Shoal by DELETING the old version (i.e. closing time brackets as of today), then ADDING the new version (i.e. appending the correct version with unclosed time brackets).

Feature	X	Y	DB From	DB To	World From	World To	
Beacon	38	38	3/21	*	3/8	*	
Shoal	23	32	5/16	9/1	5/2	8/18	(altered)
Shoal	15	21	9/1	*	8/18	*	
Hazard	56	34	3/21	9/1	3/3	8/18	(deleted)
Bell	44	28	8/30	9/1	8/8	8/18	(deleted)
Buoy	45	46	9/1	*	9/8	*	(added)
Light	12	18	9/1	*	8/18	*	(added)

9/5 transactions: ALTER Light and ADD Hazard.

Feature	X	Y	DB From	DB To	World From	World To	
Beacon	38	38	3/21	*	3/8	*	
Shoal	23	32	5/16	9/1	5/2	8/18	
Shoal	15	21	9/1	*	8/18	*	
Hazard	56	34	3/21	9/1	3/3	8/18	
Bell	44	28	8/30	9/1	8/8	8/18	
Buoy	45	46	9/1	*	9/8	*	
Light	12	18	9/1	9/5	8/18	8/18	(corrected)
Light	18	18	9/5	*	8/18	*	
Hazard	56	34	9/5	*	9/4	*	(restored)

database and world time. New tuples are added by appending them to the relation. Tuples are deleted by amending their time stamps. Tuples are altered by 'deleting' the current version, then appending the new (Table 5.4). The result is large tables that are accessed by matching tuple times to query times. Quite evidently, response time slows as the database grows; even present-tense query response time suffers from database growth.

A second tuple-based method (Ariav, 1986) orders tuples within each table and thereby violates the relational model's requirements that tuples be unordered (Table 5.5). Tuple ordering preserves object identities across time by time-slicing tables and requiring that tuples be sequenced identically in each time slice. In other words, if the table's time slices were stacked, all the tuples representing a single object's evolution would be vertically aligned.

Table 5.5 HYDRO using the Ariav (1986) method. Blanks separate time slices and are added for readability alone. A '–' signifies that the tuple is unchanged in this time slice; '#' indicates that the tuple's entity does not exist in this time slice.

DB time	Feature	X	Y	World time	
3/21	Beacon	38	38	3/8	
5/16	Shoal	23	32	5/2	
3/21	Hazard	56	34	3/3	
8/30	Bell	44	28	8/8	
–	–	–	–	–	
9/1	Shoal	15	21	8/18	(altered)
#	#	#	#	#	(deleted)
#	#	#	#	#	(deleted)
9/1	Buoy	45	46	9/8	(expected)
9/1	Light	12	18	8/18	(added)
–	–	–	–	–	
9/5	Hazard	56	34	9/4	(restored)
#	#	#	#	#	
–	–	–	–	–	
9/5	Light	18	18	8/18	(corrected)

This approach avoids using the surrogate keys that most researchers recommend to maintain object identities through time. However, the method is not without appeal; it is minimally redundant at the tuple level and conceptually simple.

A final tuple-level versioning method (Lum *et al.*, 1984) provides temporal capabilities without affecting present-tense performance. This is achieved by segregating present-tense data from historical data, connecting object versions via 'history chains' (Table 5.6).

Table 5.6 HYDRO using the Lum et al. (1984) method. A '#' means that the tuple's entity did not exist during that time interval; '' signifies a null pointer.*

Key	World time	Feature	X	Y	DB time	Next	
Present-tense:							
1	3/8	Beacon	38	38	3/21	*	
7	8/18	Shoal	15	21	9/1	2	(altered)
11	9/4	Hazard	56	34	9/5	8	(restored)
9	9/1	Bell	#	#	8/18	4	(deleted)
5	9/8	Buoy	45	46	9/1	*	(expected)
10	8/18	Light	18	18	9/5	6	(corrected)
History:							
2	5/2	Shoal	23	32	5/16	*	
8	8/18	#	#	#	9/1	3	(deleted)
3	3/3	Hazard	56	34	3/21	*	
4	8/8	Bell	44	28	8/30	*	
6	8/18	Light	12	18	9/1	*	(added)

A history chain is a list of tuples sorted with time decreasing so the most current information is also the most accessible. The rest of the chains are stored in a 'ghost' table that shares the schema of the present-tense table and intermixes the history chains of all its entities. Because of the shared schemas, superseded tuples can be accessed either by 'walking' a chain or by standard relational operators. Similarly, 'future chains' sorted with time increasing can be attached to the present-tense relation and housed in a second ghost table. Lum *et al.* suggest boosting the access rates of ghost tables with indices whose design follows that used for current data access.

Attribute-level versioning

To minimize redundancy and improve expressiveness, some researchers advocate attribute-level versioning. Attribute-level versioning requires variable-length fields of complex domain to hold lists of time-stamped attribute versions. Clifford (in Clifford and Tansel, 1985) stores time-value pairs of temporal attributes sequentially, allowing each pair to supersede its predecessor. Tansel (also in Clifford and Tansel, 1985) brackets his time intervals with start and stop times. Gadia's (1986) method, with the most painstaking recordkeeping, is demonstrated in Table 5.7.

Gadia supports time-stamping both attributes and surrogates. By time-stamping a surrogate, this method expresses lifespan better than any other constructs. Exhaustive time-stamping also avoids the tuple-based problem of tuples broken into unmatched versions within or across tables, as are Shoal, Light, and Hazard. Attribute-level versioning also prevents the misalignment that occurs when a record spans more than one table but an update to that record occurs in only one table.

Table 5.7 HYDRO using the Gadia (1986) method.

<FEA,W_FR,W_TO> D_FR,D_TO>	<X,W_FR,W_TO D_FR,D_TO>	<Y,W_FR,W_TO D_TO>
<Beacon,3/28,*,3/21,*>	<38,3/8,*,3/21,*>	<38,3/8,*,3/21,*>
<Shoal,5/2,*,5/16,*>	<23,5/2,8/18,5/16,9/1>	<32,5/2,8/18,5/16,9/1>
	<15,8/18,*,9/1,*>	<21,8/18,*,9/1,*>
<Light,3/3,*,3/21,*>	<56,3/3,8/18,3/21,9/1>	<34,3/3,8/18,3/21,9/1>
	<#,8/18,9/5,9/1,9/5>	<#,8/18,9/5,9/1,9/5>
	<56,9/5,*,9/5*>	<34,9/5,*,9/5,*>
<Bell,8/8,8/18,8/30,9/1>	<44,8/8,8/18,8/30,9/1>	<28,8/8,8/18,8/30,9/1>
<Buoy,9/8,*,9/1,*>	<45,9/8,*9/1,*>	<46,9/8,*,9/1,*>
<Hazard,8/18,*,9/1,*>	<18,8/18,*,9/1,9/5>	<18,8/18,*,9/1,*>
	<12,8/18,*,9/5,*>	

Comparison of versioning methods

Having examined the various relational methods of describing temporal data, some generalizations are possible.

Relation-level temporality is highly redundant and obscures individual object histories. However, it is conceptually simple and relates well to the current practice of generating database snapshots.

Tuple-level temporality permits better temporal resolution than the relation-based approach and at lower storage costs. Additionally, most of the relational theories and algebras apply. However, problems with expressiveness, consistency, and joins across time do exist.

Attribute-level temporality is obviously compact but requires alternate relational algebras to manage. Ferg (1985) notes that attribute-versioning shares a problem of transposed files: retrieval is fast on a single attribute but poor for a complete record, particularly one that spans several tables. To access a record as of a given date, the system must first rebuild it by retrieving each of its appropriate attributes.

Selecting a relational temporal database design

It is possible to derive tuple- and attribute-based representations one from the other using existing operators (such as 'unpack') designed to create normal-form relations from those of non-first-normal form (Ahn, 1986). However, since cost makes frequent conversion undesirable, Ahn develops a formula to compare versioning methods for a given application based on the number and size of static and dynamic attributes, the average number of updates per tuple, and the average number of attributes modified in a single update operation.

Aside from choosing a versioning strategy, a designer must choose how to order object versions within the file. As outlined by Dadum *et al.* (1984), one can represent each new version in its entirety or describe how it differs from its previous or next version. In addition, supersessions can be 'forward-' or 'backward-oriented', depending on whether they amend a past or current state. Evidently, the method chosen must depend on expected query traffic (Kent, 1982; Rotem and Segev, 1987; Langran, 1988).

Temporal operations and constraints

Codd (1981) defines a data model to encompass organization, operations, and constraints. Experiments with organization dominate the literature, as evidenced by the preceding discussion. This section treats the subject of operations and constraints on temporal data.

Operations. Temporal relational operators must extend three fundamental relational operators—Project, Select and Join—to treat the extra data dimension. Project (which accesses a column or set of columns) and Select (which accesses a row or set of rows) are readily altered by defining temporal modifiers (Ahn and Snodgrass, 1986). Comparable access along the time dimension might be defined by a Time Slice command (Clifford and Croker, 1987). But a temporal Join is problematic. Ariav (1986, p.525) defines this operator as 'the placement of two cubic extensions next to each other and the creation of a new cubic extension with a combined set of attributes, matched objects, and matched time values'. Clifford and Croker (1987) define a Time Join to join a single selected time slice of two tables. In both cases, the procedure is computationally intense and sensitive to mismatches and irregularity.

New operators could be useful. Segev and Shoshani (1987) suggest three candidates: aggregation (e.g. to tally timber sales per month for the past ten years); accummulation (e.g. to measure average annual percent change over a ten-year period); and application of a function (e.g. to interpolate an intermediate object version between two stored object versions). In addition, Ariav (1986) proposes a function to replay a sequence of events and Tansel (1987) discusses statistical operators. According to Ahn (1986), current relational languages could support aggregation but not accumulation, application of functions, replays, or statistics.

Three temporal query languages, each with a limited set of constructs, have been developed: TQuel (Snodgrass, 1987), LEGOL (Jones *et al.*, 1979), and TOSQL (Ariav, 1986). Both LEGOL and TQuel require start and end time stamps to bracket periods of tuple validity. LEGOL describes set membership using WHILE, WHERE, and DURING modifiers, while TQuel adds WHEN and WHERE to standard Ingres Quel commands. In addition, experimental query preprocessors or 'time specialists' have been developed by Overmyer and Stonebraker (1982) and Kahn and Gorry (1977).

Constraints. A basic temporal database premise is that updates do not overwrite existing information. Eliminating the rewrites of 'garbage collection' and integrating system backup, rollback, and recovery into the data model, helps protect data integrity. However, the logical constraints that enforce atemporal data integrity must be rethought.

Systems whose temporal topology is linear can constrain data to have only one active version per moment, which is preceded and succeeded by single active versions. The relational model's uniqueness assumption is in harmony with this constraint (Ariav, 1986). Chronological ordering also helps enforce this constraint (Kahn and Gorry, 1977), although ordering of tuples violates the relational model (Codd, 1981).

Constraints on completeness, density, and isomorphism have been suggested by Ariav (1986) and Clifford and Warren (1983). Temporal completeness means the database describes an enterprise's development from beginning to end; temporal density forbids gaps in the data; and temporal isomorphism prescribes a stored order and pace of events similar to those of the modelled environment.

One way to enforce logical constraints is to define a 'typed' temporal relation to specify object granularities, lifespans, types, and regularity (Segev and Shoshani, 1987). Clifford and Croker (1987) suggest that lifespans be specified for both tuples and attributes to describe their 'active' periods. Then, constraints could prohibit an attribute from holding a value during any moment not contained in its own and its tuple's lifespans.

Kung (1985) suggests three tests of database consistency. First, each database state must satisfy a set of static constraints, e.g., 'the values of soundings are below sea level and the soundings are located in the water' or 'each feature has a unique coordinate set'. Second, each sequence of database operations must satisfy a set of temporal constraints, e.g., 'lighthouses do not move' or 'once discredited, an information source cannot be used'. Finally, each operation is described by a legal precondition and postcondition and (optionally) a prerequisite database history, e.g., 'if the information source has not been discredited and the location has never held a lighthouse, then add the feature to the database, after which it will have a unique set of coordinates'.

Evidently, ensuring the correctness and consistency of temporal information is not straightforward. Structural and semantic expectations must be defined, and some redundancy must exist to permit cross-checking. Organization directly impacts the ease of operations and the effectiveness of consistency checks. For this reason, the three components of a data model—organization, operations, and constraints—should be examined in a more coherent way than has normally been the case.

Performance enhancement

No definitive benchmarks exist for temporal databases. However, the sheer bulk of temporal data forewarns performance problems. Ahn and Snodgrass (1986) did develop and benchmark a prototype temporal database management system, and found that while development was fairly simple, database size increased monotonically as it was updated. This suggests that a partitioning scheme be part of any temporal database design. Ahn (1986) also suggests using B-trees, dynamic hashing, extendible hashing, and grid files to improve access time.

Temporal database directories pose another potential performance problem. As these directories expand, they must reference both past and current data, and their pointers can change at each update operation. For these reasons, Copeland (1982) suggests that the current tradition of having record-level resolution within directories could be impractical. He proposes that directories reference disk blocks instead, with records clustered and time-sorted within blocks. This innovation would reduce both size and pointer readjustments for most directories.

Barriers to implementation

The ideas and technologies reviewed above are obviously exploitable for GIS. But one crucial question is what portions of a spatiotemporal database can be supported within the confines of the relational model. Spatial data alone tax relational databases because of relatively slow response time; adding another dimension would only aggravate this problem. A reasonable approach is to extend the popular strategy of segregating the aspatial (attribute) from the spatial (geometric and topological) data, keeping spatial data in special structures while storing attribute information in a relational database. The spatiotemporal corollary to this practice is to store both attributes and attribute change in a relational database, and to develop special structures to trace spatial evolution. The space-time composite approach follows this strategy.

Even this conservative plan has some risks, however. Temporal relational databases are understood to some degree, but unresolved issues remain, few implementations exist, a designer must choose from many organization options, new operators and constraints are required, a query language must be developed or extended, and performance must be boosted to acceptable levels. In short, the works discussed here provide a basis for design but do not provide a recipe for design.

Rudimentary tools and components now exist with which to assemble a temporal GIS. However, a definitive blueprint does not, and creation of a single blueprint to satisfy all needs is unlikely. Ultimately, many of the issues surrounding databases for temporal GIS must be resolved individually. The problem of constant identity, discussed earlier, is one such case. For each application, an entity has an essence that, if changed, causes it to become a different entity altogether; other changes merely result in a new version of the same entity. These essences must be defined prior to temporal database implementation.

Other application-specific aspects of design include how to cluster data on disk in accord with expected query traffic, partitioning large databases into subsections so performance does not suffer, and determining at what point data should be relegated to slower storage locations. The next chapter expands on these implementation issues.

Designing a digital system to replace an analog one always forces an organization to scrutinize its standard procedures and occasionally has the fortunate effect of modernizing the procedures themselves as it modernizes the techniques that perform them. As organizations implement temporal systems, they will be forced to reflect both on their data and on their role as guardians of the past.

6

Implementation issues

The conceptual model and temporal database alternatives presented earlier are fundamental components of a temporal GIS. But how to implement these constructs in a functioning GIS is problematic. Without careful planning, the massive data volumes of a temporal GIS will interfere with its usefulness and usability.[10] Throughout a system's lifespan, its designers, implementers, and managers must examine the needs of specific applications carefully and cluster data for maximum responsiveness. The system must cope with data volumes by carefully establishing appropriate temporal resolution, by retiring older data so the database does not grow indefinitely, and by partitioning data into units of workable size. Finally, the designer must examine ways to detect errors using known principles of quality control. This discussion uses the terms 'clustering', 'volume control', and 'quality control' to describe these three important facets of temporal GIS implementation.

Clustering

The discussion up to this point seems to imply that all temporal GIS needs can be met by a single monolithic design. But in reality, how data are clustered in storage hinges on the data itself and the expected query traffic. Spatiotemporal data may be predominantly spatial or temporal in form. In addition, spatiotemporal queries may request information that is predominantly spatial or temporal. To understand the implications of these variations, some background is essential.

Current hardware constraints

Presently, most stored data reside on magnetic disk. A critical objective in

---

[10] I have substituted the terms usefulness and usability for their more familiar but colloquial corollaries, functionality and user-friendliness.

database design today is to minimize the number of disk accesses that the system must make because the slow rate of retrieval from remote memory negates most processing efficiencies in main memory. Frank (1988) elaborates: one disk access requires approximately 30 milliseconds[11] and can involve 512 to several thousand bytes (depending on the system); in contrast, an operation within main memory requires approximately 0.1 microsecond[12] and involves generally four bytes (depending on word size). The difference in speed would be 300 000 to 1 if the same amount of data were involved in the two procedures. Even allowing for the different amounts of data involved, the difference in speed between disk accesses and main-memory operations easily exceeds 10 000 to 1.

Today's constraints show no sign of fading; rather, the gap between storage retrieval and memory operations is presently widening. Stonebraker (1988), when speculating on future trends in database design, envisioned single-processor capabilities of $MIPS = 2^{(year - 1984)}$. In 1990 a single chip is capable of 32 MIPS.[13] Stonebraker envisions no commensurate improvement in storage access rates. Storage costs are shrinking, but some of the newer methods (e.g. optical disk) are as slow or slower than magnetic disk.

Understanding the mechanics of data retrieval is essential to understanding system design needs. To satisfy requests for information, a system must first locate where the desired data are stored. Then the system retrieves the data by bringing them into main memory from storage. But the data are not simply plucked from storage and transferred to memory. Within storage, all data are clustered into 'pages' or 'buckets' whose size is defined by the system designer based on the hardware and software involved. Clustering ideally anticipates the groupings in which data will be requested. To transfer data to or from memory, an entire bucket is transferred, even if the desired data occupy a fraction of that bucket. If the requested data occupy fractions of many buckets, all of those buckets must be retrieved in their entirety.

For this reason, strategic clustering is critical to acceptable system performance. Some aspects of GIS operations—for example, quality checks and global searches—are unaffected by clustering because they operate on the complete database, meaning that algorithms can be designed to take data in the order in which they come. Many of the ODYSSEY system's algorithms operate in this way, and the dataflow literature of computer science addresses the broader methodological aspects of these practices (see, for example, Bentley, 1980; Dennis, 1980; Carlson, 1985; Veen, 1986). However, some GIS functions, of necessity, will access only a small subset of the database.

---

[11]One millisecond = a thousandth of a second.
[12]One microsecond = a millionth of a second.
[13]Million instructions per second.

These functions are often responses to *ad hoc* queries. While *ad hoc* queries are by no means the sole or even the most important GIS function, they generally occur interactively, which makes sluggish response entirely unacceptable. For this reason, system designers must make every attempt to speed query processing.

The many faces of spatiotemporality

The realities of physical data management cited above mean that clustering of data on disk has major effects on the user's perception of system performance because that perception often is based on how speedily it responds to *ad hoc* requests for information. Rotem and Segev (1987), in considering physical organization of temporal databases, conclude that no single structure dominates another because suitability depends on the data themselves and on the access patterns (see also Kent, 1982). It is worthwhile to address these two components in turn.

Imagine mapping spatiotemporal data to a three-dimensional phase space defined by one time and two space dimensions.[14] It is soon apparent that 'spatiotemporal' is not a precise term when applied to data forms and groupings. Within the 'spatiotemporal' class, we may see horizontal planes of data in phase space that represent the snapshot sequences described in Chapter 3 (Figure 6.1a). Alternately, we may see vertical columns of data with minimal horizontal coherence (Figure 6.1b). Or we may see a mixture of the two, where the data in phase space have coherence in both horizontal and vertical directions (Figure 6.1c). In Sintons's terms, these data fix time, space, and nothing, respectively.

Retaining the visual image of spatiotemporal data in phase space, imagine now that a query has the effect of illuminating only the data in phase space that are required for its response. Some queries illuminate planes of data, others illuminate columns, and still others illuminate mixes of spatial and temporal data that can resemble jellyfish or multi-canopied forests. The configuration of data that are illuminated does not necessarily relate to the configuration of data that exist in the phase space, but ideally there is correspondence. In other words it would be possible to illuminate planes in a columnar representation or columns in a planar representation, but in both cases the illuminated data supply a sparse and possibly questionable picture of reality.

---

[14] Physicists use the term 'phase space' to describe a hypothetical space that is constructed to have as many axes as needed to define the state of a given substance or system. This conceptual term is useful for conceptual discussions. In Chapter 7 I substitute the term 'data space' for 'phase space' as the discussion turns to implementation.

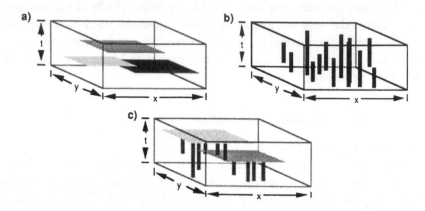

Figure 6.1 Spatiotemporal data in a three-dimensional phase space. (a) A snapshot time sequence occupies planes in phase space. (b) Object histories occupy columns of phase space with minimal horizontal coherence. (c) Spatial data with incremental changes have both horizontal and vertical coherence.

Figure 6.2 shows a selection of 'illuminated queries' in an abbreviated phase space with one spatial dimension eliminated. Each graphic in the figure illustrates the location of the data sought by a hypothetical query in a phase space comprised of one space and one time axis. Table 6.1 is a generalized listing of the types of queries that a spatiotemporal system could encounter.

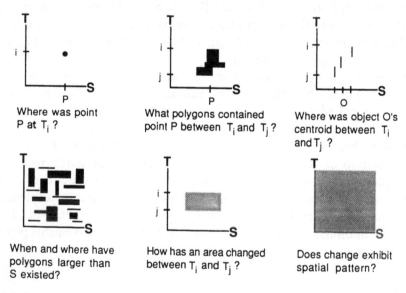

Figure 6.2 Examples of spatiotemporal queries. The data that respond to a query are distributed in a phase space comprised of one time and two space dimensions. Illustrations show the hypothetical locations of the data being sought in a cross section of that phase space.

Table 6.1 General spatiotemporal queries. These queries do not correspond to the illustrations of Figure 6.2 but rather are a generalized set of potential queries for a temporal GIS to treat.

1	Examine a feature's lifespan.
2	Examine a single time slice.
3	Examine a feature's lifespan; when the feature meets some criteria, examine its time slice.
4	Examine a single time slice; examine the lifespans of features meeting some criteria.
5	Examine the lifespans of all features.
6	Examine all time slices.

A temporal GIS faces a broad spectrum of possible data configurations and queries, all under the auspices of 'spatiotemporal'. Considering that data are transferred from storage to memory in buckets, and that each bucket retrieved is costly in terms of overall processing speed, it becomes evident that data should be clustered strategically into buckets based on anticipated retrievals.

The different characteristics of spatiotemporal systems implies that a monolithic clustering scheme may not be entirely satisfactory. A single clustering strategy can produce equally good results for all applications only by compromise. To produce the best possible performance for a given application means clustering the data to meet its specific needs. Proximity in space and time is but one indication of how data should be clustered. Within the broad generality, implementation details vary from application to application.

Because the appropriate clustering of spatiotemporal data depends on the groupings in which they will be accessed, the critical question is how the spatiotemporal information will be used by different applications. The answers are quite diverse. Of the users hypothesized in Chapter 1's scenarios, many would not even call their system a 'GIS'; they would call it an 'information system' instead because its spatial elements or analytic capabilities are relatively few. Likewise, some potential users would not consider their systems to be 'temporal' because analysis of time-dependent information is infrequent or casual. Finally, an application may refer most frequently to ranges or classes of the attribute dimension. Regardless, when one dimension dominates another, data can be clustered accordingly to improve system performance.

Dimensional dominance defined

I have coined the term 'dimensional dominance' to describe the variations within spatiotemporal systems (Langran, 1988). A spatiotemporal system is merely a system whose phase space includes spatial *and* temporal axes.

Within that broad class of systems are subclasses. Data can be configured in phase space in a space-dominant, time-dominant, or spatiotemporal pattern. Similarly, queries to the stored data can also be space-dominant, time-dominant, or spatiotemporal. These subgroupings are useful in identifying the precise nature of an application with respect to its data-processing requirements.

Figure 6.3 depicts the dimensional dominance of query-processing in various spatiotemporal applications. As with most broad generalizations, these classifications could easily change with the particulars of an application. Likewise, some shifting would occur if we were to consider instead the dimensional dominance of the data used in each of Figure 6.3's applications.

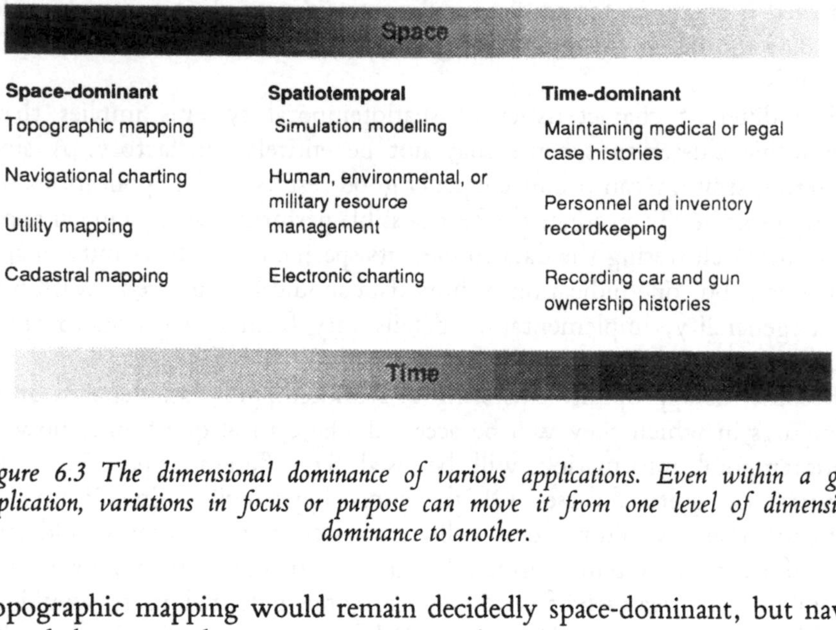

Space		
Space-dominant	**Spatiotemporal**	**Time-dominant**
Topographic mapping	Simulation modelling	Maintaining medical or legal case histories
Navigational charting	Human, environmental, or military resource	
Utility mapping	management	Personnel and inventory recordkeeping
Cadastral mapping	Electronic charting	Recording car and gun ownership histories
Time		

Figure 6.3 The dimensional dominance of various applications. Even within a given application, variations in focus or purpose can move it from one level of dimensional dominance to another.

Topographic mapping would remain decidedly space-dominant, but navigational charting and (in some instances) cadastral mapping would shift to the spatiotemporal subclass. Navigational charts are continually amended by incremental Notice to Mariners updates, as discussed in Chapter 2. Permits and licences could provide similar incremental updates to land-use mapping (Dangermond and Freedman, 1984; Vrana, 1989). As noted earlier, however, query traffic is critical to clustering, which in turn affects system performance; for that reason, I weight anticipated query traffic more heavily in my classification.

Space-dominant applications.   Map and chart producers generally are interested in present-tense data and most of their operations are limited to the spatial domain. Despite the space-dominance, the past has value to map and chart production, and a past tense does exist in the form of past map editions or incremental updates. Chart producers are liable for accidents caused by false chart information. A system that can trace past chart and database states provides some protection from unreasonable lawsuits, particularly if the lawsuit concerns errors in digital chart data disseminated by an agency. Access to the past also provides clues to the future. The work of Rhind *et al.* (1983) that attempts to forecast rates of change and apply them to production is one such instance. The future has other production value as well. Navigational chart producers often receive early notice of planned changes to navigation aids, port facilities, etc. A future tense allows them to post anticipated changes, which are absorbed into the present-tense database as time advances. Thus, production mapping and charting can be considered space-dominant but not atemporal. Other applications whose queries are similarly space-dominant are transportation, cadastral, and utility mapping. A reasonable approach to such systems is to optimize for spatial access but to provide adequate roadways into the past and future.

Time-dominant applications. Just as few space-dominant applications are wholly atemporal, few time-dominant applications can be considered entirely aspatial. Arguably, most applications are enriched by access to their spatial components. For example, medical recordkeeping, personnel tracking, and sales are often cited in the computer-science literature as temporal database candidates. But if a regional hospital's database could evaluate the medical histories of its region's residents by their location, it could lead to a better understanding of locational factors of disease. If a corporation traced where, as well as when, its employees go when they quit, it could attempt to understand the factors that draw them elsewhere. Finally, spatiotemporal sales figures are obviously useful to firms interested in purchasing cycles and patterns. While all these applications would be optimized for aspatial present-tense queries, their information systems should also support spatiotemporal analysis.

Spatiotemporal applications. The most intriguing spatiotemporal applications favor neither space nor time. These generally involve regional processes or changes in land use or land cover. Some spatiotemporal data are derived from real-time sensors or simulation. Examples include electronic navigation charts that track the vehicle's progress (e.g. Cooke, 1985; Langran and Clawson, 1986; French, 1987; White, 1987); environmental monitoring or modelling systems that study the dispersion, diffusion, or mixing of air, water, pollutants, or sediments (e.g. Samson and Small, 1984; Carmichael and Peters, 1984; Burrough *et al.*, 1988); and management systems for natural, human, or military resources where analysts might wish to gain a longitudinal perspective on a region or to play out given scenarios (e.g. Hoyt, 1970; Berry, 1974; Hagerstrand, 1974; Cebrian, 1983; Garner,

1983; Burrough, 1986; Enslin *et al.*, 1987; Maffini and Saxton, 1987; Fifield, 1987). Because such systems must be fluent in both space and time, their design is quite challenging.

Clustering considerations

Understanding the emphasis of an application is but the first step in spatio-temporal system design. The next step is to tailor the data's organization to the needs of its application, which returns us to clustering. One reason that clustering is vital to a temporal GIS is that it must accommodate queries that request all data in a range or class of values. These queries are termed 'range queries' by Knuth (1973) and 'orthogonal range queries' by Bentley and Friedman (1979).

Simple queries reference but a single data record. In one common retrieval scenario, the system finds the requested record by consulting a small index in main memory that refers the system to a segment of a larger index in storage. The system retrieves that portion of the index, identifies the storage location of the desired data, then retrieves the data. By this means, any data record can be retrieved in two disk accesses, but range queries are less easily processed because they must locate and retrieve a set of records based on value. If the requested records are not clustered in data buckets, many trips to storage ensue.

Chapter 7 discusses range queries for temporal GIS in depth. However, we can assume at this stage that spatial and temporal range queries are critical to a spatiotemporal system and defer discussion of the attribute dimensions to the next chapter. A spatial range query asks: 'what lies within a given region defined by x and y ranges?' A temporal range query asks: 'what is the history of a single entity over a range of time?' Clustering of spatial neighbors expedites spatial range queries; clustering of temporal neighbors expedites temporal range queries. Unfortunately, everything cannot be clustered with everything else.

More data dimensions mean more clustering options. One-dimensional data tend to have some natural order to exploit in the storage scheme, for example alphabetic or numeric order, but multi-dimensional data do not fall neatly into one order or another — no single natural order exists (see Goodchild and Grandfield, 1983). That is why many possible gridded data groupings exist, as shown in Figure 4.1.

Consider the tangle of one-dimensional yarn that would result from unravelling a two-dimensional knit scarf. If that scarf were a geographic region, how would we organize the stream of digital data so that, given a location on the scarf, its corresponding linear segments are easily referenced? Adding a third dimension, be it spatial or temporal, makes matters worse. A

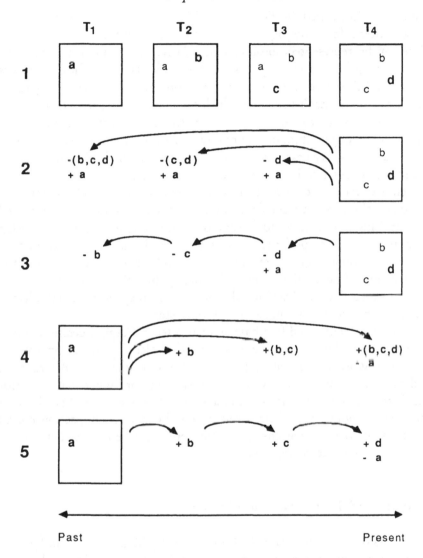

Figure 6.4 Methods of ordering temporal data. (1) Time slices stored as snapshots, (2) backward-oriented amendments to a base state, (3) backward-oriented cumulative amendments, (4) forward-oriented amendments to the base state, (5) forward-oriented cumulative amendments.

related three-dimensional analogy is to organize a strand of yarn so that it is possible to reconstruct the cloud of wool from which it was spun.

Frank (1988) discusses the problems of clustering spatial data, which relate to the lack of natural order in two-dimensional phenomena. Time's single dimension does have natural order, but several representational options are available. Dadum *et al.* (1984) and Ahn (1986) describe the possibilities;

Dadum's are illustrated in Figure 6.4. No single correct choice exists, although criteria for choosing one approach over another do.

Time can be represented absolutely or relatively. An absolute representation is a snapshot series (e.g. Figure 6.4, line 1), an option that was examined and discarded in Chapter 3. To represent time relatively requires that a base state be established from which amendments depart. Dadum *et al.* describe a base state to be the oldest or the most current data, which amendments modify; a third option not addressed by Dadum *et al.* but noted in Lum *et al.* (1984) is a base state with both forward- and backward-oriented amendments to describe both past and future tenses.

Which base state to select depends on the application. A likely construct is to designate the most current data available as a base state, with outdated data comprising a past tense and forecast data comprising a future tense. Map or chart producers may prefer the base state to be the date of the last map or chart edition produced in that area, which would make a future tense of the uncharted amendments that occurred since publication. Still others, for example historical or archaeological researchers, may elect to establish a particular past date being studied as the base state.

How amendements modify a base state is a second design decision. Fully superseding amendments are the strategy of choice for most applications because they work for all data types, nominal and otherwise. Alternately, amendments can supersede base-state information or, when data are not nominal, the amendments can accumulate from the base state. The allure of cumulative amendments is their compactness; however, compaction comes at a price, since the value at any given moment must be reconstructed from the base state and interceding values.

Case studies of dimensional dominance

This discussion on dimensional dominance highlights how differently applications will use a temporal GIS. We can legitimately ask whether it is possible to devise a single optimum design to meet the needs of all applications. Many atemporal GIS designs exist; is it reasonable to expect a single temporal GIS design to be adopted by all applications? A modular approach could improve the likelihood. If temporal GIS tools are adjustable to various levels of dimensional dominance, a designer who is aware of an applicant's use of space and time can tailor generic methods accordingly.

If the need for adjustable system tools is apparent, the precise form of these adjustments and tools is not. The question, then, is how to adapt simple data structures to different levels of dimensional dominance. This section considers three cases of designing for dimensional dominance in an attempt to shed light on appropriate procedures. Chapter 8 further demonstrates these concepts when comparing prospective data access methods.

I selected the three case studies presented here for a variety of reasons. The first, a temporal grid, is a simple construct that provides a conceptually simple illustration of clustering options. The second case presents design options for temporal relational databases (all of which were reviewed in Chapter 5). These database designs were not developed with dimensional dominance in mind. Rather, they represent different researchers' interpret-

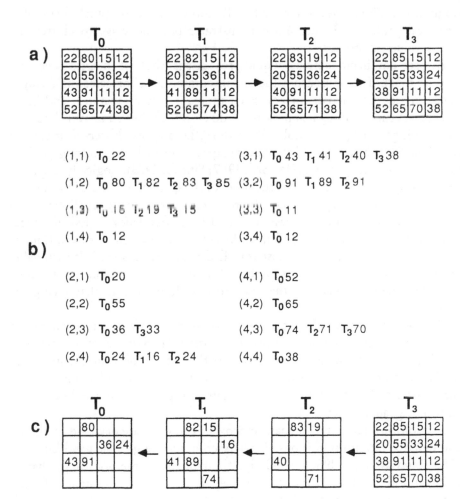

Figure 6.5 Methods of storing a temporal grid. (a) A space-dominant grid stored as a snapshot sequence; (b) a time-dominant grid stored as lists of cell values; (c) a hybrid spatio-temporal arrangement with the base state stored in space-dominant form with historical data stored in time-dominant form.

ations concerning temporal access needs. The fact that they evince different levels of dimensional dominance is evidence of the concept's generality. The final case examines nautical charting, where queries are space-dominant but the form of the data is time-dominant. I chose this example because incongruency in the dimensional dominance of queries and data is troublesome but also common in geographic applications, since a temporal land information system would share this situation.

Temporal grids. A grid, the simplest geographic data structure, provides a suitable vehicle to introduce methods of clustering for dimensional dominance. Figure 4.3 (in Chapter 4) shows the conceptual form of a temporal grid, and Figure 6.5 demonstrates possible ways to cluster the grid's values in storage.

A space-dominant approach would provide snapshots of the grid over time (Figure 6.5a). This approach is the one in common use for detecting changes between sequent Landsat images. Each image is stored separately and in its entirety. A time-dominant approach would cluster all the values over time held by a single grid cell (Figure 6.5b). The IMGRID software clusters all attributes of gridded data by storing all attributes together for each cell, although it does not use variable-length lists (Harvard, 1987; Sinton, 1978). Logically, a spatiotemporal approach might cluster subcubes of the space-time cube. However, an alternate method may produce better results for many spatiotemporal applications. By storing the base state and other frequency periods in space-dominant form, then storing all amendments in time-dominant form (Figure 6.5c) the ranges lying in the most-travelled paths are clustered. If each value changes at each time slice, this method does not improve upon the space- or time-dominant alternatives; the assumption, however, is that change is asymmetric.

Temporal relational databases. A second interesting example of dimensional dominance arises from the aspatial database literature reviewed in Chapter 5. Because this literature is aspatial, 'state-dominance' substitutes for 'space-dominance' because in both cases, it is snapshots of single states in time that are of primary importance. Figure 6.6 provides an overview of the contrasting approaches.

A traditional temporal database approach is to generate database snapshots and describe interim transitions via transaction logs (e.g. Abida and Lindsay, 1980; Lindsay *et al.*, 1986). Similarly, the table-based methods (e.g. that of Clifford and Warren, 1983) are relatively state- or space-dominant (Figure 6.6a) — a change to any field in a table results in a new table. However, both these methods become untenable if the database is large or if temporal depth exceeds several states. Space-dominant storage such as this also presents a challenge to longitudinal analysis.

If time is represented as an attribute, as in the space-time composite, attribute-based methods are time-dominant because a single field clusters all versions of the attribute stored in that field. Figure 6.6c sketches an attribute-based method where each time-varying field of a tuple holds a variable-length list of values and effective dates. Ferg (1985) ascribes the same problems to attribute-based approaches as to transposed files in general: columns of data (i.e. one attribute value for each tuple) are easy to retrieve but retrieving rows of data (i.e. all attribute values for a single tuple) is relatively slow. Because the attribute-based methods also provide the highest level of temporal detail with the most clustering, they are adaptable to time-dominant needs.

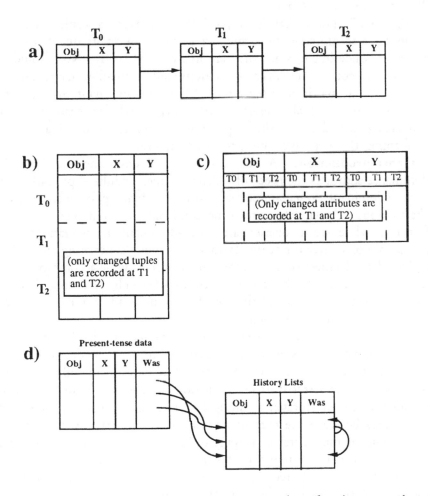

Figure 6.6 Methods of storing a temporal relation. (a) Snapshots of a relation as it changes with time; (b) changes are stored as superseding tuples in the relation; (c) changes are stored as superseding attributes; (d) changes are stored as superseding tuples, with historical data stored in linked lists in a separate table.

Several tuple-based representations exist and each places a different emphasis on space relative to time. Those that maintain transaction data with tables (e.g. Snodgrass and Ahn, 1985; Ariav, 1986) address spatio-temporal needs in a way similar to clustered space-time cubes, because all space and time data are grouped together and are equally accessible or in-accessible. Figure 6.6b sketches such a method, where all versions of tuples are stored together in a single table. Conversely, the Lum *et al.* (1984) method (shown in Figure 6.6d) stores a space-dominant base state with time-dominant amendments and addresses spatiotemporal needs in a way similar to the compromise spatiotemporal method described previously (and illustrated in Figure 6.5c).

Ahn (1987) notes that it is possible to convert a tuple-based to an attribute-based representation, and vice versa; however, frequent transform-ations are undesirable, since they are likely to be costly. Nor should it be difficult to convert between space-dominant, time-dominant, attribute-dominant, and hybrid representations.[15] If a time-dominant analysis is to be performed on a small subset of data organized for space-dominance, converting the data to the appropriate dimensional dominance before performing the analysis is one option to explore.

Chart production.   A final case study of designing for dimensional dominance addresses the needs of a chart producer. Unlike applications with a space-dominant present tense and a time-dominant past tense, many of a chart producer's updates are time-dominant amendments to existing features, while most queries are space-dominant investigations of the appearance of an area (or of the database's contents concerning an area) as of a given date. To meet space-dominant query needs while accommodating the data's temporal depth, the space-time composite can be recomposed periodic-ally to provide a snapshot of the world at meaningful moments by filtering change data from the representation.

For a chart producer, meaningful moments are likely to be chart publication dates. A system could be designed to store the data contained on a published chart (or in a disseminated chart dataset) in fully recomposed form, possibly in faster storage. Transitions between two editions could be retained in slower storage, awaiting the eventuality that they are needed. Transitions between the current edition and the known world would be retained in the main database (Figure 6.7). This support of space-dominance in a spatiotemporal system does not expand data volumes as much as one might think, since the recomposed 'chart' is actually a table of pointers to the same objects used by the space-time composite.

---

[15]Nyerges (1989) suggests that variation in dimensional dominance is an important consideration in a modelling exercise.

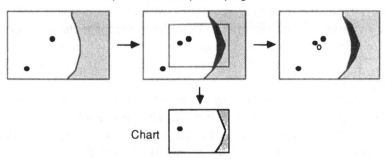

Figure 6.7 A space-dominant strategy for chart production. The space-time composite is recomposed and stored for each published chart edition. Transitions between recomposed editions are stored in slow storage, while changes since the last edition are stored online.

Summary of clustering

This discussion by no means exhausts the topic of clustering. Clustering remains an intriguing problem that is associated with multi-dimensional data of all kinds. The discussion did, however, describe the concept of dimensional dominance to help balance the needs of an application against representational realities.

While a single spatiotemporal database design is unlikely to be optimal for all applications, case-by-case design is equally undesirable. A better approach is to identify unifying characteristics among various spatiotemporal solutions, which can serve as guidelines in subsequent database and software design. Common logic and data structures unify and strengthen development efforts and permit communication and cross-fertilization of ideas. Without general rules to guide system designs into common channels, there is no means to ensure that all contingencies are met, and by grouping applications into a few classes whose functional requirements are understood, only a few spatiotemporal data designs may effectively meet all user needs.

But how different would these data designs need to be? Is a class-by-class (rather than case-by-case) solution also impractical? Spatiotemporal applications would appear to be more demanding technically than space- or time-dominant applications. It is reasonable to expect the technical solutions for space- and time-dominant applications to be subsets of the spatiotemporal solution. However, by considering the three classes separately, we avoid underestimating their differences and in that way highlight their importance to data structuring and system performance.

Volume control

The most troubling technical problem involving a temporal GIS is the appalling volume of data required to describe geographic change over time. Aside from its initial bulk, a temporal database grows inexorably and indefinitely as time progresses. In contrast, an atemporal database could remain relatively constant in size.

The amount of data involved in a temporal database permits even small storage-saving measures to have major impacts on overall data volume. All possible opportunities to control database size should be explored. Because storage volume and processing speed are fundamental design trade-offs, individual designers must judge if a size-reduction scheme is feasible for a specific application based on whether the scheme substitutes unacceptable processing overhead to alleviate storage volume.

The space-time composite employed here incorporates an important space-saving measure: it stores differences over time rather than snapshots that duplicate unchanged data. This practice is a fundamental means of controlling the expansion of a spatiotemporal database while at the same time being far more expressive and versatile than methods that store snapshots. Other ways to control data volume include carefully selecting an appropriate temporal resolution, partitioning the data into epochs, and establishing retirement guidelines so the most active data are also the most accessible.

Temporal resolution

The temporal and spatial resolutions of the data treated by a system have a major impact on data volume. Spatial resolution has been defined as 'the minimum difference between two independently measured or computed values that can be distinguished by the measurement or analytical method being considered or used' (Morrison, 1988, p. 30). This definition applies well to temporal resolution.

Spatial resolution is limited to the coordinate precision or other spatial addressing scheme, while temporal resolution is limited by the precision of the time stamps. Ideally, actual resolution corresponds to coordinate and time stamp resolution, but in reality, the collection resolution of data can lag or lead its stored resolution, with misleading or ineffectual results (Chrisman, 1984b). For example, a vector database that records vertices in decimal degrees to the ten-thousandths (e.g. 0.0001) can store a resolution of approximately 11 metres near the equator. But if the original data were collected from Landsat 1, sensor resolution and processing error would produce a maximum data resolution coarser than 80 metres, making the storage units quite misleading.

Similarly, temporal resolution can be misrepresented by its measurement units. Data collected at one-hour intervals could inadvertently be stored in a database whose finest temporal measurement is one day. Unless data are processed to produce one representative value, this situation would result in confusing and apparently duplicative data. A time stamp of one-day resolution is likewise inappropriate for a system to which updates are applied only once a year, since the added precision is wasteful and possibly misleading. As noted by Goodchild (1982), coordinate storage in spatial databases can encompass as much as 90% of overall storage cost; if time stamps are considered to be temporal coordinates, this proportion can become even more overwhelming.

In light of the above, it seems altogether undesirable to establish a temporal resolution that exceeds an application's needs, but extenuating circumstances may apply. Bauer (1984) argues that frugality in data collection is a false economy because collecting too little means the entire investment is wasted. If data are collected at one resolution for a specific application, then a later application needs a finer resolution in the same area, collecting the finer resolution originally (given the foresight) would have been far less expensive. Bouillé (1978) and Mark (1979) represent opposing views in this dilemma. Bouillé envisions a phenomena-based representation that encodes all existing entities and relationships because such a construct is versatile and possibly produces better results because it models reality. Both Bouillé and Bauer believe that comprehensive datasets will produce long-term, overall economies (see also Beard's 1987 discussion of a single-detailed database). Mark, however, suggests an application-based approach, where specific databases are developed for precise sets of requirements. An application-based approach is less versatile but also less complex because it eliminates the overhead of unnecessary data and software for specific applications.

The GIS community has only recently begun to face the dilemmas involved in general-purpose data sets. Just as ad hoc GIS dominated the field until the 1980s, data were often collected and used by one organization for one purpose, then discarded when that purpose was served. The expense and difficulty of collecting accurate geographic data has strengthened the position of Bouillé, Bauer, and Beard concerning general-purpose databases; however, only individual institutions can judge whether the general-purpose approach meets their specific needs.

The need to tie database resolution to collection rate and user needs is evident. If the two differ, a compromise must be struck, but how to develop an efficient scheme that permits data of varying temporal resolutions to coexist is more problematic. For example, a stream may be sampled at one-hour intervals for 24 hours, once a month. Is the appropriate temporal resolution for this database an hour, a day, or a month? This particular dilemma can be solved by treating the monthly sampling as a 24-field record. Many design decisions must occur case by case; however, each must weigh the added overhead of a finer timestamp against the possible loss of valuable detail.

Global partitioning

Event-oriented data representation (such as the space-time composite) combined with appropriate data resolution are instrumental in controlling database size, but in theory, a temporal database grows indefinitely, since nothing is deleted. It may be necessary in some cases to impose artificial breaks in the data to improve manageability.

GISs that cover large areas normally partition the data somehow. A global subdivision of the area into non-overlapping tiles is quite common. In that case, the entire area treated by the GIS is divided spatially into subunits (Figure 6.8a). Global tiles may relate to logical units of data, such as map sheets (strategy used by the U.S. Geological Survey), coordinate units (e.g. 1° square cells), or other meaningful subdivisions of space. The extent of global tiles may be based partially on hardware constraints. Larger tiles mean fewer partitions and less disruption of the data; smaller tiles mean less overhead in file transfer and processing. Problems associated with partitions are treated at length in Chapters 7 and 8 in their discussions of local partitioning methods.

Just as a GIS covering large regions may employ spatial partitions to make its task more manageable, temporally deep systems may require temporal partitions at a global level. Unless some horizon is introduced (over which

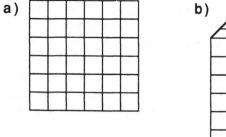 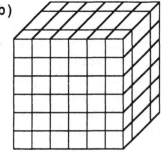

Figure 6.8 Global partitioning. (a) Spatial tiles; (b) spatial tiles divided into temporal epochs. Neither tiles or epochs are necessarily at regular intervals, since they may relate to logical subdivisions of space or time, or even hardware capacities.

objects pass to become less active) the potentially infinite lifespans of objects in the space-time composite could cause databases to grow uncontrollably. One means is to establish temporal partitions—epochs—that compare to tiles and that provide an opportunity to recompose the representation periodically. Epochs need not be at regular temporal intervals, just as spatial partitions can be irregular.

Using a space-time cube as a vehicle, Figure 6.8b illustrates how global tiles and epochs could subdivide space-time. Overall dimensions of the cube would correspond to a system's lifespan and spatial coverage; subcube dimensions would correspond to the tile and epoch sizes established for the system. Just as with tiles, selecting the appropriate epoch size is crucial. Too-small epochs would boost the time needed to respond to queries that span epochs. Too-large epochs would swell the data because of the high degree of decomposition, and also boost processing time because of the number of objects to be processed.

As illustrated in **Figure 4.5**, object lifespans are theoretically infinite. Only by imposing a horizon of some sort can the representation be recomposed into a new set of greatest common spatiotemporal units and thereby eliminate inactive objects. Epochs provide this device (Figure 6.9). Whether to employ epochs and what interval to employ relates more to the amount of

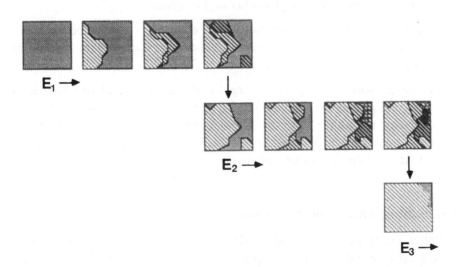

Figure 6.9 Recomposing objects at epochs. The illustration shows the gradual decomposition of a space-time composite at progressive time slices, then its recomposition at epochs, which are denoted by an 'E'. The only change occurring to this coverage is the urban area encroaching on its rural neighbour. Changes are composited within epochs to describe greatest common spatiotemporal units (i.e., polygons with histories distinct from their neighbours). At the end of each epoch, the objects are recomposed to become the initial state of the following epoch.

geometric change and the nature of expected queries than to the overall system lifespan. Highly dynamic or temporally deep databases are good candidates for epochs, although the decision rests with the designer.

The benefit of using epochs is to avoid the difficulties that a highly decomposed representation can pose. As described in Chapter 3, the space-time composite stores the greatest common spatiotemporal units, i.e. objects with coherent attribute histories that are distinct from those of their neighbors. The more spatiotemporal units required to produce a single spatial unit (i.e. a unit whose attributes at a given time slice are distinct from those of its neighbors), the more storage and processing overhead are involved. However, the cost of aggregating greatest common units is linear.

Temporal operations that span epochs require that the space-time composites of the epochs be combined to create a temporary space-time composite comprised of the greatest common spatiotemporal units of all epochs involved. Clearly, the fewer epochs a query spans, the better, since the polygon-overlay procedures involved entail log linear processing time. Factors affecting whether and how to implement epochs include the following.

- What is the average or median ratio of greatest spatial objects (at selected times) to greatest common spatiotemporal objects? Does the number of objects pose problems to the system?

- How commonly will operations span epochs? Is slower performance for epoch-spanning operations acceptable? Can older epochs be relegated to slower storage?

The last question is the key. Among the characteristics of an epoch is its ability to recapitulate a self-contained period. That trait can be used to advantage by relegating older or less referenced epochs to slower or less expensive storage. A GIS repository is likely to need a retirement policy to prevent its database from overexpansion.

Establishing retirement guidelines

Without controls, a temporal database will gradually outgrow its host system because older data are superseded but not deleted. Before storage overflows, performance becomes increasingly more sluggish as time passes and the database grows. If a database strictly adheres to a 'no deletions' policy, it must adopt some strategy for retiring less active data to more remote or lower-cost storage locations.

Today's standard database designs include archives and archival procedures. However, because an atemporal system's purpose in maintaining an archive is for backup and recovery in the case of error or hardware failure, its archives use different data structures and access procedures from the rest of the system (generally, snapshots and transaction logs are stored sequentially). Because, ideally, the archives of a temporal database use the same data structures and access procedures, I have coined the term 'retirement' as a less specific but equally descriptive alternative to 'archiving'.

At minimum, retirement guidelines should establish two levels of past: near and distant. The near past is retained in fast storage for easy referrals; the distant past is retired to archives or stored on slower media. The dividing line between the two pasts occurs where frequent references are no longer made, when newly entered data are fully checked and approved, when retroactive changes are unlikely, when map editions covering an area are complete, or using such criteria as listed above.

The structure of the retired data should depend on how frequently data will be accessed, how quick the response to accesses should be, and how compactly the data must be stored. As mentioned previously, epochs are conceptually neat storage bundles. If retired epochs use the identical structures to their more current counterparts, the same query and processing procedures can be used for data of all ages. If this convenience is less important than compactness, some overhead can be subtracted by storing retired data in a sequential or otherwise compacted form.

Summary of volume control

The change-only representation of a space-time composite is a fundamental means of controlling the volume of spatiotemporal databases. Other options include selecting an appropriate temporal resolution for the application; imposing a system of global temporal partitions (called epochs); and establishing retirement guidelines for data past their prime. Each option involves trade-offs. Tying a database's temporal resolution too tightly to a specific application could necessitate recapturing the data at a higher resolution for another application. Epochs subdivide data to improve manageability in a manner similar to global tiling systems, but the discontinuities at epoch borders increase complexity and processing overhead. Retirement schemes seem inevitable for databases that are continually growing, but retirement age and retirement formats are somewhat specific to individual applications. In each case, the designer must weigh the benefits against the costs.

Quality control

The preceding discussions in this chapter relate to practical issues of data structuring, since they can have major impacts on system responsiveness.

This final section addresses an issue that can impact system responses: data quality.

All data processing applications have a propensity for error. A temporal system must re-examine the procedures used for quality control in atemporal systems. The Task Force on Digital Cartographic Data Standards has identified five components of data quality: lineage, positional accuracy, attribute accuracy, logical consistency, and completeness.

Lineage

Lineage describes the sources from which data are captured and the methods used in the capture process. Lineage supplies a chronicle in database time of the events that have shaped the data from when it entered the system. It reports an initial world-time state, then chronicles sources, capture methods, transformations, and revisions; however, lineage does not describe the different world-time states or versions through which the data passes. The standard was defined for atemporal systems; only the lineage of current versions is traced because all others are assumed to be deleted.

While the bulk of the chronicle is clocked in database time, the standard states that the original source date should be in world time, if possible. In other words, the initial date describes when the object appeared as it is stored, while subsequent dates describe the data processing procedures performed upon it. This mixture of world and database times is not a particular problem in atemporal systems, but a cleaner separation is desirable in temporal systems.

Temporal systems offer new options for tracing the lineage of data. Separating world and database time keeps records conceptually clean. A reasonable approach is to maintain lineage entirely in database time; world time is naturally clocked by the temporal design of the system.

Completeness

Completeness defines the consistency with which selection, generalization, and categorization criteria were applied to data. Many cartographic procedures are strongly rule-based. It is particularly important to gauge the consistency of procedures that perform scale change, since misapplication of such procedures can misrepresent otherwise acceptable data. Consistency does not necessarily imply uniformity; it is conventional to apply more stringent selection strategies to densely featured areas of maps than to the sparser areas so all areas are legible but none are bare when displayed. In the absence of display constraints, however, completeness is limited by the data themselves.

For temporal GIS, we cannot assume that data collection is continual, which means that the distribution of recorded changes may not correspond to the distribution of actual changes. Ariav (1986) and Clifford and Warren (1983) define three forms of temporal completeness:

- data describe an enterprise's development comprehensively from beginning to end,
- the order and pace of recorded events are similar to those of physical environment,
- no gaps occur in the data.

It may be feasible for the data to describe an enterprise comprehensively. However, some applications will find it impossible to obtain data that continuously describe the order and pace of events in the physical world. Sources of incremental updates may be an integral part of the mapping process, as with Notice to Mariners, Notice to Airmen, and the Ordnance Survey's draft maps with surveyors' notes (Rhind *et al.*, 1983). Other untapped sources also exist; the permit process of a building or land-development office is one possible source of transaction-based data for land-use mapping (Vrana, 1989).

A less stringent definition of temporal completeness than stated above may be preferable. For example, this chapter recommends many guidelines for treating spatiotemporal information (e.g. clustering measures, standard data resolutions, standard epochs, and established retirement rules). A test for completeness could simply ascertain whether relevant guidelines are applied uniformly. As noted by Chrisman (1983), following the space-time composite approach means there is only one map to check for completeness.

Logical consistency

Logical consistency describes the fidelity of relationships encoded in the data structure and can be enforced by designing a set of structural constraints that describe correct conditions in the data. For example, the topological model in popular use today requires that chains intersect at nodes and that cycles of nodes or chains be consistent around polygons. Structural constraints for temporal geometric data represented in a space-time composite would be identical those described above.

Normally, logical consistency is checked using internal evidence based on some redundancy in the data (White, 1975, 1978; Chrisman, 1984a). Data communications employ parity-checking schemes to ensure that lost bits are noted during the data transfer process. Accountants employ double-entry notations so that errors in one location are identified through mismatches with other, supposedly matching, figures. Surveyors employ closure tolerances to provide a basis for adjustments. The topological data structure encodes relationships redundantly so that consistency and completeness may be checked.

In the framework of the temporal GIS, we can still check the logical consistency of the topological model because the space-time composite leaves it essentially unaltered. 'Temporal topology' can be constrained so that an entity can have no more than one previous and one next version. Separate constraints concerning normal form apply to the relational model.

In addition, when a database describes features, the relationships of features to objects should be described consistently. If a data structure records which objects belong to a feature and what features are described by an object, it should also be capable of testing that these cross-references correspond. This becomes particularly critical in temporal databases whose object-feature relationships continually change. Temporal databases that do not employ a feature level should require that all objects have at least one attribute set at all times ('background' and 'nothing' are legal members of this set).

Positional and attribute accuracy

Semantic constraints as described in Chapter 5 offer potential methods of identifying problems with positional and attribute accuracy by limiting the permissible values that an entity can hold. An implementation of Kung's (1985) three types of semantic constraints could expose many positional and attribute errors that are difficult to detect in an atemporal database.

To review, Kung prescribes static constraints to specify valid data states, temporal constraints to specify valid sequences of operations, and legal pre- and postconditions for all operations. Constraining the pre- and post-conditions of operations could be particularly fruitful in identifying positional and attribute error. For example, such constraints could identify potential positional error in data that had moved or otherwise changed unreasonably fast given their known character. While the discrepancy between two unmatched versions of a river may be due to actual fluvial change or geometric inaccuracy, proper use of temporal information can help distinguish between the two (Chrisman, 1983). A temporal database offers new possibilities in this area of quality control, since an atemporal database can compare only the new to the old data versions, while a temporal database can evaluate new data with respect to an entity's complete history.

Temporal information combined with semantic constraints could also expose data whose attributes change in unexpected or unlikely ways. The simplest way to check for attribute accuracy is to determine whether values fall within acceptable ranges. More sophistication is possible, even for atemporal GIS. For example, software can evaluate object relationships or assess the likelihood of an object changing from one state to the next. But far stricter error filters are possible in a temporal GIS. Using the historical data, a system can evaluate whether a change is likely, considering all of an object's previous states, the length of time since its previous mutation, and the periodicity and types of previous mutations.

Similarly, a temporal information processing scheme should exploit temporal structure to facilitate quality control and temporal analysis. By considering contiguous temporal neighbors to be topologically connected (as in Figure 3.3), a temporal topological data structure can be devised to link one to the next, thereby avoiding exhaustive searches through layers of time and space. Thus, just as the spatial topological data structure provides a means of navigating from an object to its neighbors in space, the corresponding temporal data structure would provide a means of navigating from a state or a version to its neighbours in time. In sum, the data would encode a higher level of information, since the connectivity of time is as important to temporal analysis as is the connectivity of space to spatial analysis.

Summary of quality control

Temporal information will assist in quality control because it offers a richer view of the enterprise that can be used to constrain the valid states it may hold. At the same time, temporal data makes the task of quality control more difficult because it adds to the data's complexity. To exploit the full power of stored data, each application should define structural and semantic constraints, and some redundancy must be incorporated in the data structure to permit cross-checking. In addition, data organization has a direct impact on the ease and effectiveness of consistency checks, which is a cogent argument for enumerating quality-control needs early in the design stage.

Summary of implementation issues

The theoretical and conceptual facets of temporal GIS design must join with the realities and limitations of technology to produce a realistic strategy for implementation. While conceptual designs enjoy a certain level of timelessness, implementation issues and their proffered solutions are firmly rooted in the constraints of contemporary technology. Clustering, volume, and error are the most apparent barriers to overcome for temporal GIS to become a reality. However, this may change. As noted by Chrisman (1984), storage compression at cost of software complexity is an implementation issue that changing technology has outgrown. Thus, to a far greater extent than conceptual problems, implementation issues are temporal because they change with time.

7

Accessing spatiotemporal data

The space-time composite permits us to process the three dimensions of space and time with only two axes by treating temporality as an attribute of spatial objects. Each object references a variable-length list comprised of attribute sets and validity dates that express the changes that the object undergoes. But treating all geographic change via attributes places heavy demands on attribute-processing procedures and on the linkages between spatial and attribute data. This is a serious objection because many of today's GIS segregate spatial and attribute data, either in distinct data structures or in separate databases altogether. Those that do not separate data use a relational model for all data. In both cases, the effect of cross-referencing spatial and attribute data is a negative one. Segregation in particular affects spatio-temporal procedures as envisioned here because operations must somehow straddle the two protocols.

An alternative is to describe attribute change in the attribute data, as is the current plan, but also to describe spatial change in the spatial data so it is possible to identify which objects are needed at a given time without referencing the attribute database. Many multidimensional data structuring methods exist to improve access to multidimensional data spaces. One or more of these methods could conceivably be used to access spatial objects according to location and lifespan.

This chapter explores potential methods of spatiotemporal data access. The discussion begins by identifying the basic data groupings to be accessed and outlines the procedures needed to access them. I then examine existing multidimensional data structures and consider their suitability for spatio-temporal systems. I define a taxonomy that separates access methods according to their operation. I then describe the data structures in operational terms, since some operational strategies face standard sets of strengths and weaknesses. In Chapter 8, I use taxonomy as a basis for selecting contrasting approaches to compare.

Access patterns in a temporal GIS

Organizing a three-dimensional data space is a challenging goal. Chapter 6 describes different ways to cluster multidimensional data, whose multiple dimensions provide them with no single natural order. The appropriate treatment of geographic data is predicated upon access patterns, i.e. the cumulative access paths used by an application. Since appropriate clustering in storage can speed access rates considerably, I begin this analysis by revisiting the question of access pattern.

Data access in a GIS is not guided solely by *ad hoc* queries. Rhind and Green, (1988) list six basic GIS procedures (data input, manipulation, retrieval, analysis, display, and management) that have subcomponents unrelated to query processing. Nonetheless, query processing is among the more demanding of GIS procedures. This is due, in part, to the fact that other procedures can be designed to operate on data in any order and therefore present few access problems. It is also partly due to the interactive nature of query processing, which causes users to judge performance quite harshly and to consider sluggishness intolerable. For this reason, the focus here is on access patterns that are evidenced by the query traffic of an application.

Query types

All data access procedures operate in a search space that lies within data space. Data space is a k-dimensional space whose axes are defined by the query's components and delimited by the domain of those components. Search space is a k-dimensional subspace within the data space that is defined by the way the query qualifies its components. To illustrate using an aspatial example, a building of given height, color, and material has a zero-dimensional search space located in three-dimensional data space (Figure 7.1). A class of buildings of specified color and material whose heights fall in a given range can be represented by a one-dimensional search space in the given three-dimensional data space.

Assuming that the three fundamental components of geographic information are location, attributes, and time, the global data space of a temporal GIS lies within a set of axes that describe these components. A query to a temporal GIS defines a search space within the data space by constraining location, attributes, and times to points or segments along each axis.

Table 6.1 lists a set of fundamental spatiotemporal queries. At the root of that listing are four primitive query types.

1. Simple temporal query, i.e. what is the state of a feature at time *t*?
2. Temporal range query, i.e. what happens to a feature over a given period?
3. Simple spatiotemporal query, i.e. what is the state of a region at time *t*?
4. Spatiotemporal range query, i.e. what happens to a region over a period?

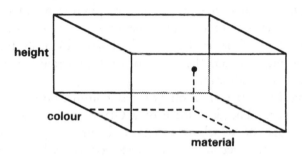

Figure 7.1 *A three-dimensional data space defined by the attributes height, colour, and material. The desired data have been narrowly specified to reduce the search space to a single point.*

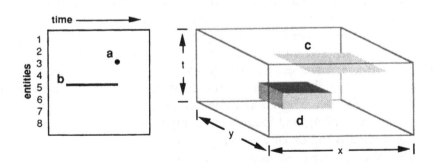

Figure 7.2 *The data and search spaces of four primitive geographic queries. (a) A simple temporal query references a single point in a data space defined by attributes. (b) A temporal range query references a vector within attribute data space. (c) A simple spatiotemporal query references a plane within a geometric data space. (d) A spatiotemporal range query references a cube within which the desired data lie.*

Queries two, three, and four are 'orthogonal range queries' (as defined by Bentley and Friedman, 1979), also known simply as range queries (as defined by Knuth, 1973) because each describes its search space to be a range within *k*-dimensional data space. The search spaces of the latter two queries are defined by some spatial criteria, i.e., within a radial distance of point x or line y, or in the area contained within boundary b. Referring back to Sinton's (1978) classification of representation methods, Queries One and Three fix time; Query One controls attribute (to the domain held by that feature) and Query Three controls location to the region desired. Despite the fact that time is fixed, the simple queries phrased by One and Three are distinctly products of a temporal GIS because they permit a user to specify at which time to fix. Queries Two and Four are obviously temporal because they fix nothing; instead, they control time to a range of values and thus breach Sinton's classification.

To distinguish further these query types, consider how each defines its data and search space. The data space of a GIS feature is defined by attribute dimensions, one of which is time. A feature version occupies a zero-dimensional search space in that data space because each of its attributes has a single value along one of the axes of data space (as in Figures 7.1, above, and 7.2a). A feature's temporal trajectory is marked by a range along the time axis, which defines a vector in data space because, at each location, along the time range, each feature attribute has a single value (Figure 7.2b).

The data space accessed by a spatiotemporal query differs from that of a temporal query. A temporal query defines its data space by thematic attributes and time. But a spatiotemporal query defines a geometric data space within which to model cartographic objects. The 'geometric' data space is geometric in the sense that it provides a medium for describing the measurements and relationships of points, lines, and areas. Thus, while the attribute data space resembles the physicist's phase space because it is constructed to describe multiple states of substances or systems, a geometric data space is closer to Hagerstrand's aquariums (1970), Rucker's space-time cubes (1977 and 1984), and Szego's cartographic vision of the world (1987) because it models the extent, movement, and change of entities.

If the data subspace of interest is actually a point location, the search space of a simple spatiotemporal query is also a point. Likewise, a linear subspace has a correspondingly linear search space. But in most cases the subspace is two dimensional; then a simple spatiotemporal query defines a two-dimensional search space (Figure 7.2c) and a spatiotemporal range query defines a three-dimensional search space within the geometric data space (Figure 7.2d).

A final and critical difference between temporal and spatiotemporal queries is that temporal queries always seek points within the search space, even if the search space itself is multidimensional. This is because each feature version (e.g. each tuple) has a single attribute value for each axis of data space. While I have defined temporal topology to be a chain whose nodes are birth and death and whose vertices are attribute change, these nodes and vertices are represented as discrete events in the attribute data space. In contrast, spatiotemporal queries seek points, lines, and areas within the search space's quasi-physical x, y, and t dimensions because these are the cartographic objects that the system models.

The discussion that follows requires some terminology to discriminate between objects that are points in the thematic data space and those that are points, lines, and areas in a geometric data space. I have chosen the terms 'dimensionless' and 'dimensional' to describe thematic and geometric objects, respectively. It is true that temporal thematic objects occupy a range in time and thereby acquire a dimension; however, this simple terminology facilitates all subsequent discussions.

The challenge of treating cartography's dimensional objects lies in their irregularity. Aside from having up to two spatial dimensions, cartographic objects have irregular shapes that cause them to extrude through three-dimensional search spaces and necessitate complex computation to detect intersection (Figure 7.3).

How to store these disorderly data in some orderly fashion is the focus of the analysis that follows. One option is to map cartographic objects in a higher-dimensional data space in which they are points. Nievergelt *et al.* (1984) suggest this treatment for multi-keyed attributes; it is worthwhile to examine this treatment for objects of multiple dimensions. Spatiotemporal data would require six dimensions into which an object's minimum and maximum x, y, and t values would be mapped. For the moment, however, this option is set aside in favor of examining less exotic measures that require a mere three dimensions.

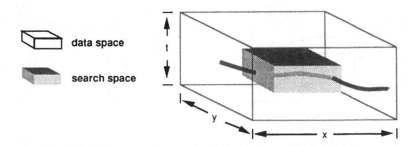

Figure 7.3 Cartographic objects extrude through spatiotemporal search spaces. The chain that snakes through the search space in this figure is monotonic with respect to time.

Procedures for query response

Data retrieval for temporal and spatiotemporal queries begins in different data spaces, which I will consider to be treated by different databases altogether. This is certainly the more common approach. Spatial and attribute data generally are segregated in separate data structures that are linked by common object and feature identifiers (e.g. ODYSSEY, Delta Map, ARC/INFO/MGE and to a lesser extent TIGRIS). Roussopoulos and Liefker (1985) argue that geometric and alphanumeric data must be maintained in separate databases so that each revolves around its optimum procedures, and so no compromises are required to adjust one database to the needs of the other. Some designers do treat both spatial and attribute information in a single relational database (e.g. Cox *et al.*, 1980; Lorie and Meier, 1984; Tuori and Moon, 1984; Dueker, 1985; Waugh and Healey, 1986; Van Roessel, 1986; Abel and Smith, 1986).

The integrated approach is not as neat in the conceptual terms of the preceding discussion. The distinct characters of the spatial and attribute data spaces cause the integrated approach to intermix a continuous data space with a discrete one. It will also be handicapped by the somewhat sluggish performance of the relational database compared to special-purpose spatial data structures. Temporal queries address the attribute data primarily, and spatiotemporal queries address the spatial data primarily (although in both cases, reference to the secondary data can occur) (Figure 7.4).

The space-time composite as defined thus far describes all temporality via a time axis in the attribute data space, which is represented as a time stamp in the attribute database. This segregation permits us to treat time aspatially and space atemporally. Changes to geometric objects spawn new objects that replace existing objects; by definition, each object in the space-time composite has a single geometric and topological description throughout an

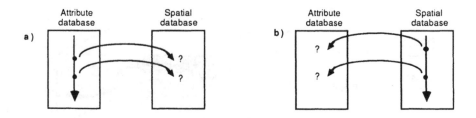

Figure 7.4 Temporal and spatiotemporal access procedures. (a) Temporal data access begins in the attribute database, although spatial qualifiers may necessitate crossing over to the spatial database. (b) Spatiotemporal data access begins in the spatial database, although reference to the attribute data may be necessary.

epoch. Using the space-time composite strategy advanced thus far, the attribute database is the clearing house for all temporal qualifiers of both temporal and spatiotemporal queries. The attribute and spatial databases are cross-referenced by object identifiers or feature identifiers (if applicable).

It is useful at this stage to step through the procedures necessary to access the search space defined by each of the four primitive queries within this segregated architecture. For the sake of simplicity, I use the relational model's term 'tuple' to describe the composite representation of an object or feature version.

Simple temporal query. A simple temporal query seeks a feature version that was current on a specific date. How the system proceeds depends on how the feature is specified. One method of specifying a feature would be for an analyst to point to its image on a graphic display and ask about another of its versions. The feature image is generated by assigning symbols to objects in the spatial database; objects and features are cross-referenced using common identifiers, which are sometimes called keys. To respond, the system uses the feature key to cross from the spatial to the attribute database, locates the pertinent information (including spatial object keys), crosses back to the spatial database, collects the appropriate objects, and displays the results. In this case, the attribute data space is defined by the two axes of feature key and time (as in Figure 7.2a, above). No geometric data space is referenced because the spatial data are sought using object keys alone.

An alternate way to specify a feature is to state the attribute values it must possess. In this case, the query defines a data space whose dimensions correspond to the attributes specified. The search space is defined by the value(s) along each axis that the feature must process. This type of query can be treated solely within the attribute database unless the qualifications

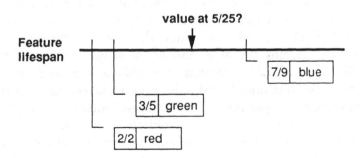

Figure 7.5 Event-driven temporality means that there is not necessarily a tuple to correspond to each time.

include reference to the feature's spatial components. Spatial qualifications include 'within a given distance of a second feature', 'intersecting a given point, line, or area', 'contained within the area bounded by . . . ', etc.

While the attribute database is likely to include some spatial information (e.g. area, perimeter, center of gravity, or minimum bounding rectangle), we cannot guarantee that the data space of a temporal query will not occasionally include geometric dimensions. As illustrated above, attribute and geometric data spaces are quite different conceptually; a geometric data space houses dimensional objects. For this reason, a temporal query with spatial qualifiers must reference two separate data spaces, entailing a search that crosses back and forth between the spatial and attribute databases. How this is managed relates to the power of the access method and the linkage between the two data spaces.

Laying aside the issue of dual data spaces, a second problem remains. Finding a feature version as of a given moment is not straightforward. A temporal database is event-driven so a tuple does not necessarily have a time value to match every *t* (Figure 7.5).

As argued previously, a feature lifespan can be considered a chain whose nodes correspond to birth and death, and whose vertices are points of attribute change. To time slice a feature is to locate the value of a point along its temporal chain based on the time stamp that equals or immediately precedes the requested time. This implies that ordering a feature's versions within storage would be helpful, just as any chain's vertices are stored in sequence.

Temporal range query. Temporal range queries are subject to the same problems as simple temporal queries, with the added difficulty of treating the temporal range. The system must locate all versions of the desired feature that existed during any part of the specified time span. The system can select all tuples where Feature = (some qualification), Time ≤ Maximum Time, and Time ≥ Minimum Time. Or, if tuples are linked in temporal order (as suggested above), the system can locate a tuple at one end of the desired time range then 'walk' through time-sorted versions until reaching the other end of the range. Because of the data's event-oriented temporality, locating the ends of the range may involve finding the times that precede and follow the actual endpoints of the range.

Spatiotemporal queries. The database literature supplies ample guidance on accessing dimensionless objects in attribute space, and the temporal database literature offers methods to access dimensionless objects in a temporal attribute space (as reviewed in Chapter 5). However, the problem remains of how to access dimensional objects in a spatiotemporal data space of the sort required by a temporal GIS.

A simple spatiotemporal query requests the current versions (as of a given date) of all objects that intersect a specified spatial range.[16] According to the strategy presented in Chapter 3, the space-time composite responds to a simple spatiotemporal query by locating the spatial objects that meet the intersection criteria, crossing to the attribute data space to reference their attributes as of the desired time, then using those attributes to identify and dissolve 'unnecessary' objects. Unnecessary objects are chains that separate polygons of like attributes, and chains or nodes with attributes of 'background' or 'null'. This last step recomposes the greatest common spatiotemporal units into greatest common spatial units for that particular query. A spatiotemporal range query differs only in that the system seeks all attribute records falling within a time range.

As the space-time composite's decomposition increases, i.e. when many spatiotemporal units comprise a single spatial unit for a particular time, the procedure outlined above becomes progressively less acceptable. The system must locate and cross-reference the attributes of all objects before they can be filtered to produce the final object set. The more objects that prove unnecessary, the greater is the method's inefficiency.

An alternate approach works for feature-based systems only. In that case, the attribute database holds feature data that describe which feature versions are current at the requested time(s). Feature records reference objects in the spatial database, which permits the system to assemble a spatial description. But this method, too, is potentially inefficient. Spatial descriptions in the attribute database are sketchy at best, which means that access will involve many uncoordinated (hence inefficient) trips to storage. Also, since several features may reference the same object, each trip to storage must stop at a checkpoint to confirm that the object being sought has not already been retrieved.

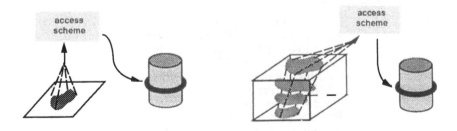

Figure 7.6 The access of spatial and spatiotemporal objects. An access scheme supplies directions to storage locations for specific objects and speeds data access.

---

[16]Either or both of the two spatial ranges can degenerate to express a point, line, or area.

Developing alternatives to these two access strategies is the focus of this analysis. Current geographic data processing procedures include methods for structuring spatial objects to speed access based on locational criteria, and many purportedly k-dimensional data structures exist that, reasonably, could serve to speed access to spatiotemporal objects based on locational and temporal criteria (Figure 7.6).

Accessing data in multidimensional space

Of the many access schemes that now exist under the title 'multi-dimensional', not all are equally so. Most were developed specifically to access dimensionless attribute data in a k-dimensional data space (see the first grouping of Table 7.1, and Six and Widmayer, 1988). Still others are not entirely suited to reference k-dimensional search spaces within the k-dimensional data space.[17] Since our goal is to access multidimensional geometric data in multidimensional search and data spaces, these procedures clearly fall short.

Table 7.1 Accessing objects of zero, one, and more dimensions in k-dimensional space.

Zero-dimensional	K-d tree (Bentley, 1975)
	K-d-b tree (Robinson, 1981)
	Multikey hashing (Rothnie and Lozano, 1974)
	Extendible hashing (Fagin *et al.*, 1979)
	Point quadtree (Finkel and Bentley, 1974)
	Multidimensional trie (Orenstein, 1982)
	Multidimensional directory (Liou and Yao, 1977)
	Log log n structure (Fries *et al.*, 1987)
	Quintary tree (Lee and Wong, 1980)
One-dimensional	Strip tree (Ballard, 1981)
K-dimensional	R-tree (Guttman, 1984)
	R + tree (Roussopoulos and Leifker, 1985)
	Packed R-tree (Faloutsos *et al.*, 1987)
	Cell tree (Gunther, 1986)
	Grid file (Nievergelt *et al.*, 1984)
	BANG file (Freeston, 1987)
	Quad-CIF tree (Kedem, 1982)
	BSP tree (Samet, 1984)
	Region quadtree (Fuchs *et al.*, 1980)
	EXCEL (Tamminen, 1981)
	Field tree (Frank, 1983)
	Pfaltz (1979)

---

[17]For example, some hashing methods purposely disperse data throughout storage, rather than clustering neighboring values, which makes range searches inefficient (Nievergelt *et al.*, 1984).

Noronha (1988), Chrisman (1990), Kleiner (1989), and Goodchild (1989) review the access methods that are suited to referencing geometric objects. Table 7.1, drawing on these comments, lists methods designed to access a multidimensional data space according to the number of object dimensions each can treat.

Accessing options for zero- and one-dimensional objects are of little use to the planned strategy of mapping three-dimensional objects in three-dimensional data space. Therefore, the next step is to examine methods that treat k-dimensional objects in k-dimensional space (Table 7.1's third grouping), since these are the methods that can be used to access spatio-temporal objects and which, potentially, could also access the attribute database.

Several writers have attempted to classify data access schemes. Nievergelt *et al* (1984) define two broad classes: a scheme can adapt to the data themselves or partition the embedding space. In the first class, the actual locations of objects in data space determine partitions. K-d trees and R-trees are representative of this approach. Notable members of the second class include grid files and quadtrees, which subdivide when a predetermined sector of data space exceeds a predetermined maximum capacity. Gunther (1986) advances a similar classification by distinguishing trees (whose decomposition is guided by the configuration of the input data) from tries (whose decomposition is preordained and is only triggered by the quantity of input data). These frameworks are quite useful conceptually but enough hybrid schemes exist to make them something less than definitive. Specifically, the quad-CIF tree associates the minimum bounding rectangles of its objects with cells of a recursively subdivided space, into which minimum bounding rectangles do not necessarily embed completely (Noronha, 1988). Other schemes that resist the trie/tree classification are the BANG file and extendible hashing.

Freeston (1987) defines a pragmatic classification of access schemes: tree structures, extendible hashing, and grid files. But where the Nievergelt framework is perhaps too conceptual, this classification is too mechanistic. The strengths and weaknesses of quadtrees and grid files, and those of R-trees and BANG files, are more similar than those of quadtrees and R-trees, or grid files and BANG files. Yet Freeston's classification results in the latter two dissimilar groupings based on common data structuring mechanics. In addition, some schemes cross Freeston's class bounds (e.g. EXCEL uses a hashed index to a set of partitions that resemble a grid file).

Noronha (1988), too, defines a classification scheme for access methods. This scheme distinguishes hierarchical versus nonhierarchical and regular versus object-oriented subclasses to highlight the major performance differences among methods. However, Noronha's goal is to examine data-structuring options, and he avoids the rigor of a taxonomy.

Chrisman (1989) addresses the dilemmas of partitioning that are reviewed later in this chapter. He highlights several important distinctions in access schemes that are incorporated in my taxonomy below. In particular, his discussion of quadtrees separates those whose nodes reference objects from those whose nodes reference object addresses only. He also separates quadtrees that store objects only at leaf nodes from those that store objects throughout the tree.

A taxonomy of access methods

A taxonomy that describes access methods based on overall approach would help clarify the options available. My taxonomy, presented below, uses terminology that departs from earlier conflicting usages. What I call 'access methods' have been interchangeably termed 'indexing' and 'partitioning' by the researchers whose work is reviewed here.[18] My taxonomy distinguishes indexing from partitioning because these two components of an access method are associated with separable functions and ramifications. In essence, a partitioning scheme subdivides data space to define how data are clustered into buckets and an indexing scheme defines the procedures a system follows to locate data given full or partial spatial, temporal, or attribute specification.

Not all indexes can be applied to all partitioning methods and a sharp line between the two procedures is occasionally difficult to draw. Nonetheless, it is useful to consider each separately because of separate literatures and distinct performance considerations.

In general, indexes are rated according to their algorithmic efficiency, which is based upon best- and worst-case measures of performance. Peucker (now Poiker, 1976) and Chrisman (1979) discuss the performance measures of cartographic algorithms. A general expression of performance for indexes is how many buckets of data must be retrieved from storage to produce a requested data item. Performance efficiency is also a concern of partitioning, but it is strongly affected by whether a given level of partitions subdivides data space into mutually exclusive or overlapping cells, and whether the partitions clip or avoid objects in the data space (Chrisman, 1990).

Indexing

Indexes are references to logical or physical addresses that provide access to stored information. An indexing scheme prescribes the strategy that a system

---

[18]Lodwick and Feuchtwanger (1987) have used the terms 'access method' and 'storage structure'.

uses to search for information when complete or partial descriptions of qualifying traits are provided. Each indexing scheme is associated with measures of efficiency that are based on the storage used and the processing time required (Bentley, 1980). Storage varies with different indexes because each uses a different structure to link related data items. Atre (1980) defines storage efficiency to be inversely related to the average number of bytes required to store a byte of raw data.

Storage efficiency is important, but modern hardware constraints (reviewed in Chapter 6) make processing efficiency a primary concern. Processing efficiency in database systems tends to be bound by how quickly the system can produce requested data from storage. Atre (1980) defines processing efficiency to be inversely proportional to the average number of physical accesses required per logical access (e.g. the number of disk accesses needed to produce the requested data record). Unnecessary disk accesses are a major concern for any data processing application, for reasons cited earlier.

In general, each entry of an index references a data bucket in mass storage. Data buckets can contain actual data records or merely addresses of records stored elsewhere. The major indexing methods are search trees, directories, and hashing. Chrisman (1990) notes the emergence in geographic information processing of a fourth indexing method, topological navigation. Figure 7.7 relates these four methods to some of their variations.

Indexing classes are not mutually exclusive. Topological navigation often supplements a search tree or hashing scheme, which provides the starting point for navigation. Directories and hashing are also sometimes used in

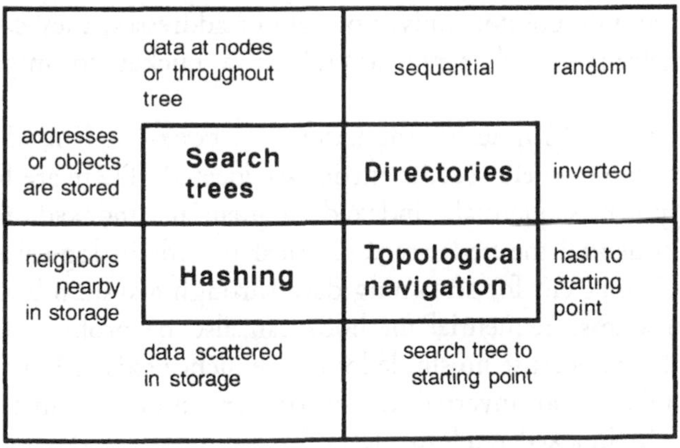

	data at nodes or throughout tree	sequential	random
addresses or objects are stored	**Search trees**	**Directories**	inverted
neighbors nearby in storage	**Hashing**	**Topological navigation**	hash to starting point
	data scattered in storage	search tree to starting point	

Figure 7.7 A typology of indexing methods. Indexing can involve one or more of the procedures listed in the inner box. The variations within each procedural class are listed in the outer box.

conjunction; the hashed value indicates which portion of the directory to retrieve from storage. Finally, data partitions can be hierarchical but not use a search tree to access nodes; a hashing scheme or directory provides entry to the hierarchically stored data.

A search tree has a defined hierarchical order with pointers between nodes. Search trees for multidimensional data reference subdivisions of space whose resolution changes systematically as the distance from the root increases. To locate data in the tree, one enters at the root node, which represents the complete region treated by the tree, and traverses the tree by selecting a path at each juncture to narrow the search space. Search trees are a highly adaptive type of data structure that originally were developed for ordering dynamic data in main memory (e.g. Peterson, 1957; Hoare, 1961; Adelson-Velskii and Landis, 1962; Clampett, 1964; Floyd, 1964; Williams, 1964). B-trees (Bayer and McCreight, 1972; also see Comer, 1979), which are used to search dynamic files on disk, followed the introduction of main-memory search trees by ten years (Fagin *et al.*, 1979).

In general, the search efficiency for trees is inversely proportional to the number of levels in the tree, since that is the number of disk accesses required to locate a given key (Knuth, 1973). Evidently, an unbalanced tree loses storage efficiency because of unnecessary depth. This measure of efficiency assumes that nodes of the tree reference unique portions of data space; the discussion on partitioning that follows includes an exception to this assumption. If the data space of nodes does overlap, the efficiency decreases by the number of levels below each overlap area because all overlapping branches of the tree must be searched. The measure of efficiency also assumes that each tree node corresponds to a separate disk bucket. However, if high-level nodes contain pointers only, not data or addresses, they can be retained in main memory or clustered several to a bucket to improve storage efficiency.

A directory works much as the index of a book—it is used as a look-up table to determine where specific entries are located. There are four directory classes: physical sequential, indexed sequential, indexed random, and inverted. Sequential methods store physical records in logical sequence and are extremely efficient for streaming data through a system but unworkable for random access. Sequential methods can also be problematic for multi-dimensional data because no single logical sequence exists. For these reasons, indexed random and inverted directories are more common in today's databases. The indexed random method maintains an index entry for every database record. The index itself is either ordered sequentially or accessed via a hashing algorithm (see below). In contrast, inverted methods build separate indexes for each attribute; for each possible value of an attribute, its index lists the locations of all records that hold that value.

All directories have similar performance efficiencies. Directories are based on the assumption that one disk access is required to read the appropriate portion of the directory and discover the bucket in which the requested data lie, then a second access to acquire that bucket. In theory, then, a directory should provide data in two disk accesses. In reality, the measure is affected by the partitioning scheme, how much of the directory resides in main memory, and the relationship of data record size to bucket size.

The purpose of a hashed index is to compute a logical or physical storage address that is a function of some components of that data. Hashing techniques are the data processor's response to our imperfect world, in which data have duplicate keys, are distributed non-uniformly in data space, or are otherwise incompatible with sequential storage. At their best, hashed indexes are the fastest of all the indexes because the address is a function of the data itself, which means the system can locate the data in a single disk access. In reality, however, 'collisions', or bucket overflows can occur when the hashing function places too many values in the same bucket. A number of remedies exist for these situations, but each adds complexity and reduces efficiency. A second problem with hashing for spatial applications is that many methods scatter data throughout storage because their goal is merely to create a uniformly distributed sequential file from the input data. In contrast, hashing approaches used for spatial information strive to preserve the spatial neighborhoods of the input data to the extent possible (e.g. Morton, 1966; Orenstein, 1982; Tamminen, 1982; Abel, 1986). Commonly, a spatial hashing procedure represents each attribute value (e.g. the x and y coordinates) as a bit string and interleaves the bits of the two strings to produce a new value that is used as an address (see Goodchild and Grandfield, 1983).

Search trees, directories, and hashing are common to all data-processing applications. Conversely, topological navigation is peculiar to dimensional applications. If data records supply pointers to neighbors, accesses within neighborhoods can use these 'topological' data as stepping-stones to navigate from one object to another. The DIME file's ARITHMICON (White, 1975), Chrisman's polygon overlay algorithm (1974), the ETAK automobile navigation system (White, 1987), and the TIGRIS engine (Herring, 1987) demonstrate topological navigation in a spatial system. Chrisman (1989) names several tessellation systems that use this principle (e.g. Brassel, 1975; Gold, 1978; Lukatela, 1987). Lum *et al.* (1984) demonstrate navigation in the aspatial context of a temporal relational database.

Marvin White (1975; 1978; 1987) has been a particular champion of the navigation approach, and was instrumental in its implementation in the Census Bureau's DIME files and TIGER system and in the ETAK electronic map system. The ETAK application uses topological navigation to traverse a road network that is topologically encoded and whose linkages are stored as physical addresses (Chrisman 1990). Because the system is assumed to be installed in a vehicle that moves continuously through the network, knowing a vehicle's past position, speed, and bearing indicates its present position (Figure 7.8a). This form of topological navigation resembles dead reckoning, where present position is derived from past position, bearing, and distance travelled.

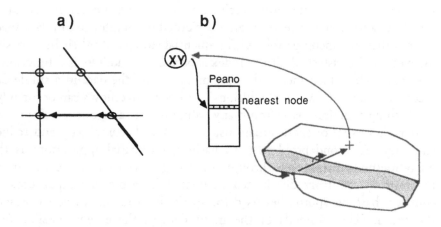

Figure 7.8 Topological navigation. (a) Navigating along chains resembles inertial, navigation systems where present position is computed from past position, bearing, and distance. (b) Navigating across space is also possible using the topological network to move from one polygon to the next.

The TIGER system uses an areal version of the navigator approach. Using a system of Morton hashing (Morton, 1966), the node nearest to the desired point is located in storage by a hashed key that is computed by interleaving the bits of the x and y, coordinates. The system uses relative position of the node to the desired point, combined with knowledge of the node's chains, polygons, and neighbors, to navigate to that point in storage (Figure 7.8b). Chrisman (1974) and Brassel (1978) discuss the specifics of this 'cross-country' journey.

Indexes have substantial impacts on performance. Search trees are highly adaptive but sensitive to depth and balance. Their search efficiency is a function of the depth of the tree. Overlap between nodes or imbalance in the tree structure degrades the efficiency, while storing high-level nodes in main memory and invoking effective balancing procedures improves it.

Directories are usually too large to retain in main memory, so they often are implemented with a preliminary indexing procedure to retrieve the required portion of the directory from disk. The preliminary index may be either a directory-to-the-directory or a hashing function. In theory, directories can locate data in only two accesses, although the partitioning scheme is critical to performance (as discussed in the next section). The price of this expedience is the overhead of the directory itself.

In contrast, hashed indexing involves virtually no storage overhead, but hashing does not guarantee a unique address, which means that algorithms must incur the processing overhead of treating collisions and overflows in storage (see Davis and Hwang, 1986). A final indexing scheme—topological navigation—is perhaps in a class by itself. Navigation is only useful with dimensional data and usually is employed as a secondary indexing scheme after the primary scheme or some other data manipulation has provided an entry point from which to navigate (Chrisman, 1990 cites dbmap of Cooke and Dawes, 1987 as an exception). However, topological navigation is a common means by which a GIS can search for data, and therefore qualifies as an index by my definition. Topological navigation entails no storage overhead because it operates on topological data, a natural resource of most systems today. Its processing can be quite efficient if data are clustered so navigation occurs largely within a single bucket. However, because navigation uses relative location, not absolute addresses, to compute an index, it is possible that it will entail the retrieval of unnecessary buckets (Chrisman, 1990).

Because of the importance of range searching to geographic applications, an indexing scheme that can find a single data item in the least number of disk accesses may not provide sufficiently rapid performance. The goal is to find ranges of data in a minimum number of disk accesses. Some clustering scheme is required (as discussed in Chapter 6) to group data and thereby optimize range queries. The dimensionality of spatial data makes spatial partitioning a logical choice for determining clusters.

Partitioning

Indexes to spatial data generally reference segments of partitioned data space that correspond to geographic areas, which are often called 'tiles'. Chapter 6 coined the term 'epochs' as the temporal corollary of tiles. To avoid confusion, I reserve these terms for global partitions that subdivide the full extent of time and space treated by a system.

Within a tile or an epoch, local partitions may subdivide the data into bucket-sized clusters. In a spatiotemporal system, these partitions can result in 'cubes' of data whose dimensions are space and time. A term does exist for cubic units of data: voxels (a contraction of volume and pixels). But voxels, like pixels, imply raster data units and not containers for vector data. For this reason, I avoid the term and instead use the more generic term 'cells' for all units created by local partitions.

Because cell size is constrained by bucket capacity, bucket size is a critical factor of any partitioning scheme. As described in Chapter 6, a system's manager or designer establishes a bucket size that is appropriate to a given application based on hardware and software considerations. A cell can hold no more objects than can be described in the maximum number of bytes per bucket. A cell that exceeds that number must be subdivided or allotted another bucket.

A wide array of partitioning strategies exist, and are discussed at some length in both Chrisman (1990) and Noronha (1988). Chrisman and Noronha both treat partitioning as integral to indexing. This discussion acknowledges the symbiotic affect of the two upon performance but discusses their characteristics separately.

Partitions can be at one level or hierarchical. They also can divide data space into regular or irregular cells. Irregular cells can overlap or not within a level. Successive levels of a regular hierarchy can nest fully or not. Figure 7.9 depicts my interpretation of how these subclasses relate.

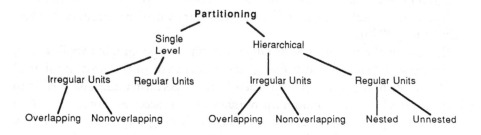

Figure 7.9 A taxonomy of multidimensional partitioning schemes.

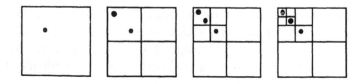

Figure 7.10 The quadtree's regular nested partitions. This demonstration shows a progression of hierarchical decomposition as new points are added. Each cell corresponds to a bucket and bucket size is set at one to provoke rapid partitioning. Bucket overflow causes a cell to subdivide into quadrants.

This discussion focuses on hierarchical partitioning schemes. The uneven spatial distribution of geographic features makes it difficult to partition at one level because a single cell size is not suitable for diverse data densities.[19] Appropriate cell size depends on data density. Cells that are too small incur excessive overhead and complexity; cells that are too large contain too many objects, causing performance problems in exhaustive searches.

In contrast, hierarchical partitions permit overcrowded cells to be subdivided into subcells so the number of objects per cell is relatively uniform. A common regular hierarchical partitioning scheme is the quadtree approach, which decomposes overflowing cells into quadrants (Figure 7.10).

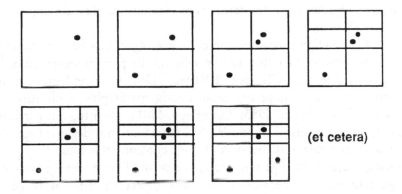

Figure 7.11 The grid file's regular nested partitions. Partitions are progressively refined as new items are inserted. Each cell corresponds to a bucket and bucket size is set at one to induce rapid refinement. A new partition bisects the overflowing cell, alternating directions.

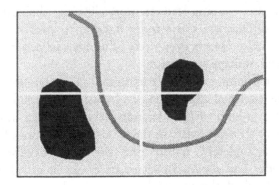

Figure 7.12 Objects clipped by regular partitions.

---

[19]Frank (1983) estimates that densely populated areas of a geographic region can have 1000 times more data than its sparser areas.

A second regular hierarchical approach is described by Nievergelt *et al.* (1984) for use in the grid file. When faced with cell overflow, the Nievergelt method bisects that cell with a new partition, alternating which axis the new partition parallels (Figure 7.11).

Figures 7.10 and 7.11 demonstrate the quadtree and grid file partitioning schemes using dimensionless data because it simplifies the illustrations. However, if data are dimensional, it is impossible to establish regular partitions that do not clip at least some objects (Chrisman, 1990) (Figure 7.12). The choice, then, is either to deal with the clipped objects or to take measures to prevent partitions from clipping objects. 'Clipped' objects are crossed by one or more cell boundaries, but not necessarily cut into new component objects. I use the term 'split' for objects that are clipped, then cut.

To deal with clipped objects, the complete object can be duplicated in all cells it intersects or the object can be split and its parts represented in intersecting cells. The quad-CIF tree of Kedem (1982) uses the former strategy; the latter strategy is more common among geographic applications (e.g. Aronson and Morehouse, 1983; Tamminen, 1981). Most regular partitioning schemes, including grid files and BANG files, were not designed specifically for dimensional data and therefore do not confront this issue.

Storing a complete object or its reference in all intersecting cells boosts storage requirements because a complete object description is generally bulkier than partial descriptions. But splitting objects boosts processing time because broken objects must somehow be zipped together before analysis (e.g. Beard and Chrisman, 1988). Remedial measures are essential; a system that cannot deal gracefully with clipped objects risks providing false responses to queries that span more than one tile (Chrisman, 1990) (Figure 7.13). In either case, clipping leads to more clipping; the new objects created by splitting or duplication cause more buckets to overflow, which refines partitions, which clip objects even more. The analysis described in Chapter 8 demonstrates this unhappy situation.

Without abandoning regular partitions, it is possible to avoid the inevitable inefficiencies of clipped objects by using hierarchical partitions and storing objects in a level of the hierarchy where they are not clipped. Two approaches to this strategy exist. In the first, the partitions of one level of the hierarchy nest completely within the partitions of the preceding (or succeeding) level. The region quadtree of Fuchs *et al.* (1980) adopts this strategy. Each level of the nested hierarchical cells stores the objects that would be clipped by the next-smaller set of partitions. But while this approach effectively avoids clipping objects, it incurs a different sort of inefficiency that relates to the indexing scheme.

The approach taken by Fuchs *et al.* uses a search-tree index to organize the

Select all polygons of size ≥ S **Select all polygons of size ≤ S**

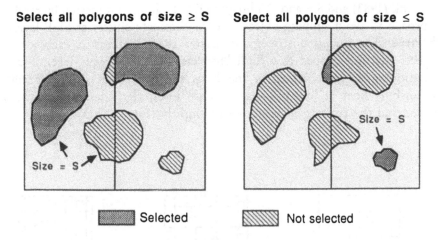

Size = S

Size = S

▒▒▒▒ Selected   ╱╱╱╱ Not selected

Figure 7.13 Objects split by regular partitions can produce false responses to queries on size.

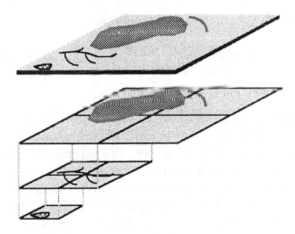

Figure 7.14 Objects stored in a regular hierarchical partitions can be stored in several tiers to avoid being split. In this case the small arc clutters the highest level of storage because it crosses a partition at the next-lower level.

hierarchical partitions. To use the index, the system enters the tree from the root node, which corresponds to the coarsest level of partitions. Objects stored at the root level are examined each time a search occurs. Objects stored at lower levels are examined each time a search passes through their node. Ideally, then, objects should be stored in the lowest nodes possible for the best efficiency. But if a small object that could fit unclipped into a lower level of the hierarchy crosses a partition, it remains to clutter a high level of the hierarchy. In Figure 7.14, the small semicircle illustrates this problem.

Frank (1983) and Six and Widmayer (1988) adopt an unnested hierarchical approach to avoid clipping objects and to eliminate clutter at high levels of the tree. The unnested strategy translates the partitions at each level so objects that are clipped at a high level can fall to a lower level. Frank's partitioning scheme, which is the basis for field trees (1983), becomes successfully finer at lower levels. At each level, the sides of the cells are halved and the entire grid is translated slightly (Figure 7.15).

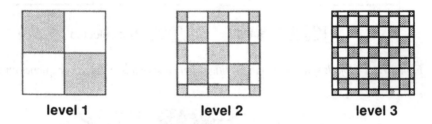

level 1   **level 2**   **level 3**

Figure 7.15 Frank's unnested partitions (1983) use regular square cells that are quartered and translated slightly at each successive level.

Frank (1983, p. 106) states that it has been proven that any object will fit unclipped into a cell with a side 16 times its maximum extension (although it may fit into smaller cells). However, this scheme is hampered by several complexities that impede the normally streamlined operations of search trees. The unnested nature of cells prevents simple comparison searches that dictate which path in the search tree corresponds to which section of search space. Also, due to the nature of translations between the partitions, an object may be unclipped at levels one and three but be clipped by level two, which makes balancing the tree somewhat challenging. In addition, because the method dictates that objects will be stored in the lowest level where they are unclipped, each search is forced to proceed to the lowest tree level. Frank's remedy is to establish minimum levels of storage for different 'types' (e.g. countries precede states, which precede counties) to permit a search to halt when it has reached the minimum level for its type.

The Six and Widmayer (1988) scheme differs from Frank's because the unnested partitions become progressively coarser at each successive level (Figure 7.16). As with Frank's strategy, clipped objects are transported to a different level of the file with different partitions. Thus, larger objects percolate to the lower levels and the smallest objects are stored in the highest level. This inversion is not a problem because Six and Widmayer use a directory rather than a search tree to index the partitions. The directory describes where partitions are placed at each level of the hierarchy.

The strong point of this scheme is its adaptive partitioning strategy that inserts a new partition only when a bucket overflows. A new partition is positioned in the vicinity of the overflow and in a location of maximum distance from partitions of all preceding layers. The weak point of this scheme is that, as with the Frank method, most searches will involve all levels of the hierachical partitions, producing poor best-case but adequate worst-case search efficiency. I selected this strategy to examine in more detail for spatiotemporal purposes and describe its mechanics at length in the next chapter.

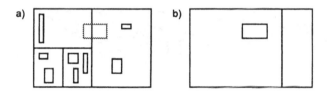

Figure 7.16 The Six and Widmayer unnested partitioning strategy inserts a new partition only when a bucket overflows. Each partition is placed to maximize its distance from the partitions of all preceding levels. (a) The primary partitions clip one rectangle. (b) The clipped rectangle is placed in a second grid file whose partitions are offset.

The offset of regular partitions does avoid clipping objects, but only at the cost of involving more levels of the hierarchy in each search. An alternate means of clipping objects is to dispense with regular partitions altogether. The irregular approach that is best represented in the literature is the R-tree (Guttman, 1984), which divides data space into overlapping subspaces built around the minimum bounding rectangles[20] of object occurrences (Figure 7.17). The strength of this approach is that objects are not split. Its weakness is that the irregular cells can overlap, so an object can belong to more than one cell.

As evidenced by its name, an R-tree employs a search-tree index to organize its partitions. To choose a path through a search tree involves examining the data space contained in each node of the lower level, then moving downward to the appropriate node. By this means, large portions of data space are discarded at each downward move to progressively narrow the

---

[20]'Minimum bounding rectangle' in its common usage (and as used here) is a misnomer. Such a rectangle minimally bounds an object while remaining orthogonal to the x and y axes.

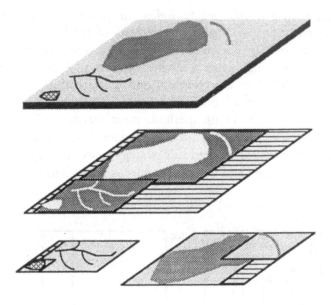

Figure 7.17 Objects stored in irregular hierarchical partitions do not need to be clipped, but cells can overlap and cause inefficiences in the search tree. To search the data space in the overlapping area of the partitioning scheme would require traversing both the cells that contain that data space.

search. If the data space represented by a node is not quite unique within its level, the search cannot be narrowed and search efficiency is no longer a function of the depth of the tree; rather, a search adds d additional bucket accesses where d is the depth of the tree below the overlap node. To seek a specific object or area within that overlapping data space, the system must traverse more than one branch of the search tree.

To avoid problems with overlap, packed R-trees (Roussopoulos and Liefker, 1985) employ more extensive heuristics to minimize overlap within the tree. The Faloutsos work demonstrates that overlap can be avoided altogether if objects are points, but that overlap cannot be avoided for dimensional objects. A second attempt to address the problem of overlap resulted in R^+ trees (Faloutsos *et al.*, 1987). The R^+ tree, when faced with an object configuration that would lead to overlapping areas of data space at high levels, simply removes the overlap by splitting the data space along the offending partition. It then represents any objects that are clipped by this artificial fissure in both branches of the tree.

A final irregular partitioning method is the cell tree of Gunther (1986). Cell trees, as the other irregular schemes, were designed specifically for dimensional data. Unlike all other schemes reviewed here, the cell tree does not partition based on minimum bounding rectangles of objects. Rather, it represents objects as polyhedra that roughly correspond to the 'greatest common units' used by geographers. Cell trees operate much as R-trees by partitioning data space to envelop the polyhedra. Just as with R-trees, a polyhedron that is clipped by the partitions is stored in both branches of the tree. Cell trees have not been tested through implementation and no performance measures are available.

Using the taxonomy

This taxonomy of access schemes helps to summarize the strengths and weaknesses of a given access scheme based on its class. Partitioning introduces particularly troubling dilemmas. Single-level partitioning schemes are simple but inefficient because they cannot adapt to uneven data densities and because searches for subcells must proceed exhaustively. Hierarchical schemes can adapt to uneven data densities and can reduce or eliminate exhaustive searches of the data space, which makes them the only reasonable approach for interactive operations.

Of the hierarchical schemes, those with regular cell units must split clipped objects, store them redundantly, or adopt a strategy that avoids clipping altogether. Those that clip objects incur added storage overhead (which the next chapter demonstrates). Those that avoid clipping are more complex and suffer diminished search efficiency because of the ambiguous placement of objects within partitions. Schemes that employ irregular cells experience a new set of problems. A designer must choose between an irregular system with overlapping cells or a system whose partitions clip objects at some level and must reference those clipped objects redundantly.

The index that an access uses also affects performance. Indexing is the main distinction between quadtrees that are traversed as search trees and line quadtrees that are accessed by hashing. Indexing also distinguishes the EXCELL scheme (Tamminen, 1981), which uses a hashed index, from a grid file (Nievergelt *et al.*, 1984), which uses a directory. Each type of index has associated costs and benefits, enhancements and trade-offs.

The usefulness of the taxonomy to this study is that it distinguishes approaches by their functional traits. To proceed, I choose distinctly different approaches and implement them on sample spatiotemporal data. Chapter 8 describes this process.

8

Comparison of access methods

Of the access methods reviewed in Chapter 7, none have been tested on spatiotemporal data. If the access procedure is comprised of indexing and partitioning, as Chapter 7 suggests, we can consider how to implement each of the two subtasks in turn using spatiotemporal data.

It would appear that indexes are less sensitive to added dimensions than partitions because an index merely defines a search procedure and provides the appropriate pathways. A number of proven multidimensional indexes can be exploited for spatiotemporal purposes. Multidimensional search trees have been demonstrated for both main memory and storage access (Bentley, 1975; Robinson, 1981, respectively). Multidimensional directories form the basis of the popular grid file (Hinrichs and Nievergelt, 1983; Nievergelt *et al.*, 1984). Three-dimensional hashing is less represented in the literature, but does not appear to pose any significant obstacles to implementation. Navigating through three-dimensional space, while presenting interesting possibilities, is likely to present minimal risk to implementation.

In contrast, partitioning three-dimensional space-time would appear to aggravate the partitioning problems discussed in Chapter 7. Partitions subdivide data space to define logical (and possibly physical) containers for objects. There is no reason to believe that an object's temporal dimension is more amenable to clipping that its spatial ones. Duplicate representation of clipped objects is equally undesirable in the temporal domain as in the spatial, and false responses to queries concerning duration would be a problem if objects are split between two partitions. Alternately, approaches that avoid clipping objects are more complex than methods that clip; how these approaches would perform using spatiotemporal data is unclear.

This chapter explores the dilemmas of spatiotemporal data partitioning by examining four contrasting approaches. Using the taxonomy of Chapter 7, I chose four hierarchical partitioning schemes of different classes. I then implemented each approach using test data and evaluated their strengths and weaknesses qualitatively.

Study goals

Use of a multidimensional access method in a spatiotemporal system offers the potential for improved performance in retrieving dimensional data. It is conceptually simple to respond to a spatiotemporal query using the space-time composite: the system references the attribute set that is current at the requested time for each object in the space-time composite, then discards unnecessary objects. But this strategy is not always efficient, since a highly decomposed representation must reference the attribute sets of many objects that ultimately are discarded.

Thus, the broad goal of this study is to develop a general strategy for more direct access to spatiotemporal data stored in a space-time composite. 'Direct' implies that not all objects are processed; only those that are needed at the requested time are sought. The opposing approach, which entails blanket processing, I will call 'exhaustive'.

In pursuit of a direct access method, I screened the major multi-dimensional partitioning approaches to find those that qualify for more stringent testing. My goals were to develop a general strategy for direct spatiotemporal data access, compare likely approaches, and improve our understanding of the mechanics of spatiotemporal access procedures. A final study goal is less applied than those that ostensibly drive it. None of the purportedly k-dimensional access methods reviewed in Chapter 7 are illustrated using more than two dimensions. The reasons for this are now obvious to me: conducting a three-dimensional evaluation and illustrating its results on a two-dimensional page is challenging. However, this practical constraint also begs the question of how the structures behave as dimensions are added. My study does not answer this question entirely, but it does provide some indication of what the answer could be.

Overall design

I designed my evaluation to be qualitative rather than quantitative, since the research is purely exploratory. I implemented each approach on small test datasets, casting myself in the role of the computer. After structuring the data according to each approach, I queried my structures using a procedure designed to estimate best- and worst-case performance. My output is exposition and graphics, since performance statistics on datasets so small would be meaningless.

A phenomenological approach such as mine offers many advantages over a programmed, computer-driven analysis, particularly at this exploratory implemented each approach using test data and evaluated their strengths and then weaknesses qualitatively.

decisions and can respond with different ones. This provides an intimate view of the trade-offs between the methods and the overall complexity and completeness of each. The next step following a study such as mine would be to focus on the more promising methods and test their performance statistically using realistic data as input and the computer as processor. A qualitative analysis permits fine-tuning and iterative enhancement of selected methods, using the performance data as feedback. However, proceeding with a quantitative analysis without first undertaking a qualitative evaluation would produce statistics to describe what happened without necessarily providing insight into why, when, or how. My evaluation is intended to produce such insights and to indicate the necessary algorithmic ingredients for computer instructions.

Functional requirements

Data structures have characteristic capabilities that make them acceptable or unacceptable for different applications. Three characteristics that are important to consider for spatiotemporal purposes are whether a structure supports range queries, whether a structure can define a search space along any of its dimensions, and whether a structure could treat dynamic data.

Chapters 6 and 7 discussed range queries and their importance to temporal GIS at some length; it is sufficient to reiterate here that performing efficient range queries with respect to x, y, and t is critical to the success of a spatio-temporal access method. Range queries delimit intervals of data space, which become the boundaries of a dimensional search space. We can expect to receive requests for:

- all versions of an entity over a given time span (i.e., in a range of t),
- all objects that intersect a line or area (or a range of x or y),
- all information that intersects a space-time cube of known dimensions (i.e. in a range of x, y, and t).

The first request retrieves the information needed to respond to a temporal range query (as defined in Chapter 7); the second collects the geometric information needed to respond to a simple spatiotemporal query; the third gathers geometric information to respond to a spatiotemporal range query.[21]

A final characteristic is difficult to gauge for spatiotemporal data. A broad distinction among data structures is how well they adapt to dynamic data, i.e. data with frequent insertions, deletions, and changing relationships. A static data structure tends to perform better on relatively stable data because

---

[21]The first of the four query types defined in Chapter 7 is not a range query and so is not discussed here.

more effort is expended on optimizing for a specific configuration of data. A dynamic data structure can restructure itself based on insertions or deletions, but its fluidity comes at the cost of deteriorating performance over time. Nievergelt *et al.* (1984) and Roussopoulos and Liefker (1985) supply broader discussions of the cost of dynamics in data structuring.

Whether a temporal GIS needs a static or dynamic approach depends on the application. A temporal GIS epitomizes dynamics, since its goal is to describe change, but the temporal data itself is relatively static. We can assume that a database's present tense will be its most dynamic part because world-time changes modify the present-tense data. However, the dynamics of temporal systems differ from the additions, deletions, and modifications of atemporal systems. Temporal system dynamics are almost solely additions, since such systems treat modifications and deletions as supersessions by clocking events in both world and database time. Lack of deletion may not be characteristic of all temporal systems, however. A system may choose to treat the time continuum's tenses discretely, in which case graduation from the present to the near past, or from the near past to the distant past, could appear to the data structure as a removal from one and an addition to the other.[22]

In summary, temporal data are not uniformly static or dynamic, and how the dynamics manifest themselves depends on an application and its implementation. Rotem and Segev (1987) argue for using a static data structure on historical portions of temporal databases, under the assumption that postactive changes will be few. The dividing line between static and dynamic data will be application-specific; conceptually, however, it should correspond with the boundary between the near and distant past, as described in Chapter 6.

The access methods that I compare in this chapter are all capable of treating dynamic data; in one case, procedures exist to improve a dynamic structure's static performance (i.e. the packed R-trees of Roussopoulos and Liefker, 1985). For that reason, I defer a decision concerning static vs. dynamic structures to individual system designers.

The four approaches

The data structures listed in Table 16's third grouping (Chapter 7) can treat range queries along any dimension and can adapt to limited data dynamics.

---

[22]Discrete tenses should not be confused with discrete change, which is a component of cartographic time. Discrete change means that versions of objects can be firmly separated by a time stamp; discrete tenses mean that a system treats an object version as a formal member of a tense rather than as moving gradually away from the present.

This leaves a large number of methods that potentially meet spatiotemporal access requirements. To select a representative subset of these methods to evaluate, I first selected three classes from the taxonomy: regular nested partitions, regular unnested partitions, and irregular partitions. Regular nested partitions are the simplest approach; they also illustrate the costs of clipping objects in three dimensions. Regular unnested partitions avoid clipping objects by providing several levels of offset regular partitions in which objects can be placed. And irregular partitions avoid clipping objects because they are formed upon the objects themselves.

Given these three classes, I selected a method to represent each: the original grid file of Nievergelt *et al.* (1984) to represent regular nested partitions, the offset grid file of Six and Widmayer (1988) to represent regular unnested partitions, and the original R-tree of Guttman (1984) to represent irregular partitions. In addition, since alternatives to clipping are important to consider, I chose to evaluate a six-dimensional grid file that avoids clipping dimensional objects by reducing them to points in six space.

The choice of R-trees is a simple one because it is the only irregular cell method in common use.[23] The choice of the grid file and offset grid file is not straightfoward, however, because several alternatives are popular among researchers.

The most obvious alternative to the grid file would have been the quadtree. Grid files and quadtrees both impose a foreordained partitioning scheme upon data space, following a trie rather than a tree strategy (Gunther, 1986). Quadtrees are a dominant presence in the GIS literature (e.g. Samet *et al.*, 1984; Chen and Peuquet, 1985; Mark and Lauzon, 1985; Davis and Hwang, 1985) and have been extended to a third dimension as octrees (e.g. Jackins and Tanimoto, 1980; Gargantini, 1982; Jackins and Tanimoto, 1983; Ayala *et al.*, 1985; Mark and Cebrian, 1986). Waugh (1986) calls them 'trendy' and Chrisman (1990) states that quadtrees now seem to have a life of their own. So many variations on quadtrees now exist that a complete commentary on the alternate configurations would entail a major digression from the discussion at hand.

Briefly, however, much of the quadtree literature pertains to pixel-based systems, which have little in common conceptually or structurally with the vector-based systems considered here (see Waugh, 1986 for a rebuttal of the pixel-based approach). The same is true of octrees, which generally manage voxels, not partitions. The earliest quadtrees (Finkel and Bentley, 1974) were intended to treat dimensionless data. Some vector-oriented quadtrees (e.g. the region quadtrees of Samet, 1984) do not take paging of secondary

---

[23]Among the GIS that use R-trees are Deltamap (Blackwell, 1987), TIGRIS (Herring, 1987), and, according to Chrisman (1990), possibly System 9.

memory into account, which limits their usefulness for database applications. These manifestations of the quadtree are easily excluded from my study.

Of the quadtrees that are functionally suited to my requirements (e.g. Fuchs *et al.*, 1980; 1983), their trie-oriented methodology makes them highly sensitive to the positioning of objects relative to the grid, an objection that also applies to grid files (Gunther, 1986). The PM quadtrees of Samet and Webber (1985) are less sensitive to grid position, but they also cannot be generalized to more than two dimensions, nor are they useful for range searching (Gunther, 1986).

For my purpose of comparing partitioning approaches, however, quadtrees are obviously poor performers. Their branching factor is 2^d for d dimensions (Gunther, 1986) and the placement of partitions is quite rigid. Chapter 7 compared the partitioning procedures of grid files and quadtrees. The grid file procedure is more adaptive and less sweeping than the quadtree method because it adds new partitions one by one (rather than two by two) and inserts new partitions, one at a time, where data are concentrated. Because partitioning is my focus, I chose the more adaptive approach to study under the assumption that it also offers the most flexibility for extension to a third dimension.

Another regular-partitioning option is the binary space partitioning (BSP) tree of Samet (1984). Like quadtrees, BSP trees divide space recursively into subspaces separated by hyperplanes. Hyperplanes correspond to interior nodes of the tree, and each partition corresponds to a leaf. As noted by Gunther (1986) however, BSP trees are typically very deep, not very dynamic, do not account for paging in secondary memory, and consequently perform poorly. Van Oosterom (1989) has extended the BSP tree to treat multi-scaled information. This extension helps an application limit how deeply into the tree it traverses, but also further limits how dynamic is the structure. It also makes generalization to three dimensions more complex.

These relatively unattractive alternatives justify the choice of a grid file to represent the regular nested class. Only two well-known methods exist in the regular unnested class: the field tree (Frank, 1983) and the offset grid file (Six and Widmayer, 1988). The choice in this case was based on the flexibility of the partitioning strategy. Field trees have predefined partitions; conversely, the offset grid file inserts partitions only as data densities warrant, although the placement of partitions is equally deterministic. Both methods seem reasonable, but the offset grid file appeared to be slightly more versatile.

Study design

Given four possible methods of accessing spatiotemporal data, I designed my study to examine how different partitioning approaches affect data accessibility. To compare the approaches, I implemented each using test data, 'queried' my implementations, then considered the problems of organizing and accessing data according to each partitioning scheme.

The sample data

Two general methods exist for testing experimental procedures. One method acquires real-world data on the assumption that a large enough dataset incorporates all possible irregularities and complexities and thus provides an adequate assessment of how well and completely the procedures perform. Naps and Singh (1986) counsel against using this method, arguing that the results of such experiments prove little beyond the procedure's ability to process those particular data. Instead, they suggest itemizing the contingencies that can occur, then custom-designing data to incorporate them. Their suggested method is to classify input data into equivalence classes, each of which tests the same set of possible conditions. By this means, a small amount of data can test system response to a comprehensive set of inputs.

I was not testing software; rather I was comparing different approaches to a problem. For that reason, I adopted a simplified version of the Naps and Singh approach. I considered the general cases that the access method would encounter and designed sample data to represent them. In addition, I did not develop code so that a computer could perform the necessary manipulations; because these implementations were ill-defined, I performed them myself as a way of defining them more clearly. This provided me with instant feedback concerning the effects of each implementation decision.

The most important cases to represent in the sample data are the different data types and the full spectrum of changes each can undergo. I developed two sample datasets to represent line and area features, respectively. Each dataset describes the data at both the object and feature levels. I treated the two data types separately for clarity; mixing them would not change the implementation methods but would make illustrating the outcome more difficult. Because points are dimensionless and therefore not truly spatiotemporal I did not illustrate their implementation here.

Features change in different ways. Dimensional features are born, die, and can move, change shape, or be reincarnated. A death causes a feature's object(s) to assume the value(s) of neighbors. Reincarnation causes a new set of objects to replace the old, even if it is identical to that of its former life, because objects are not reincarnated.[24]

The sample lines include three features: A, B, and C (Figure 8.1). Line A dies and is reincarnated. B moves at T_b to intersect C. C changes shape at T_c to intersect B a second time. Figure 8.2 shows a space-time composite of this snapshot sequence. A total of 13 chains represent Figure 8.1's changes (the thirteenth chain is Chain 1, which I count twice because it is reincarnated).

Areas undergo the same changes as lines. Figure 8.3 shows a snapshot sequence of the sample area data, which incorporate each of these changes. Area A dies and is reincarnated. B is born at T_b, then moves to intersect Area C at T_c. C changes shape at T_b.

Areas can be represented by closed sequences of chains or by rings (which reference chains).[25] My implementations use the chain as the basic processing unit. The DIME files (Cooke and Maxfield, 1967) and POLYVRT (Chrisman, 1974) were early implementations of this approach. The Canadian Geographic Information System (Tomlinson *et al.*, 1976) was the first GIS to use chains as the processing unit; other adherents include ODYSSEY, MGE, and ARC/INFO. Figure 8.4 shows the area data in a space-time composite using a chain representation. A total of 16 chains describe the snapshot sequence of Figure 8.3 (Chains 12 and 14 are reincarnated and therefore counted twice). The 'sliver' polygon that Chain 6 abuts is disregarded because it is below the tolerance size that would qualify it as an object rather than a discrepancy.

The alternative to chain representation, rings, incurs an overhead of ring/chain cross-references that are unnecessary for my purposes. To represent an area by listing its bounding chains is more compact than describing a ring apart from the chain network. Rings introduce an added layer of data: areas reference rings (in a many-to-many relationship), and rings reference chains (in a many-to-many relationship). Figure 8.5 shows the sample area data represented by rings to demonstrate the expanded storage requirements over chain representation.

A second benefit of dispensing with rings is that they tend to be more expansive than the chains that comprise them. Larger sized objects are more likely to be clipped or to cause overlap between irregular cells, causing

---

[24]The reason for not reincarnating objects is a technical one: it limits the amount of variable-length data that the system must treat. By defining objects to have one life only, their records need have only one field each for birth and death.

[25]I am using the terms 'area', 'ring', and 'polygon' as defined by the National Standard. An area is a bounded continuous two-dimensional object that may or may not include its boundary; a ring is a closed sequence of nonintersecting chains that does not include the interior area; and a polygon is an outer ring, an interior area, and zero or more nonintersecting, unnested inner rings (Morrison, 1988, p. 27–28).

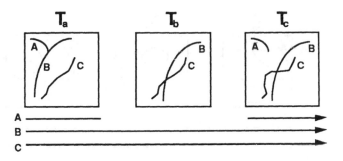

Figure 8.1 Sample line features. Line A dies and is reincarnated, B moves to intersect C, and C changes shape and also changes where it intersects B.

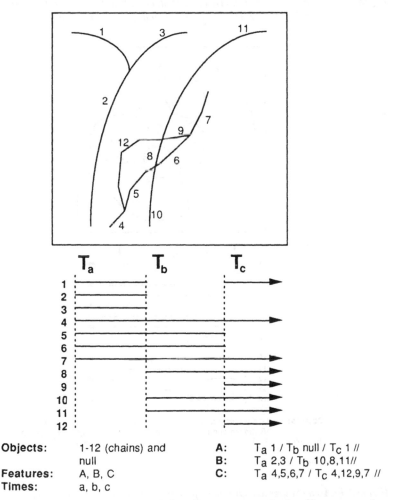

Objects:	1-12 (chains) and null	A:	T_a 1 / T_b null / T_c 1 //
Features:	A, B, C	B:	T_a 2,3 / T_b 10,8,11//
Times:	a, b, c	C:	T_a 4,5,6,7 / T_c 4,12,9,7 //

Figure 8.2 Lines in a space-composite. A total of 13 chains represent the three features over three time slices (the thirteenth chain is Chain 1, which is reincarnated and counted twice).

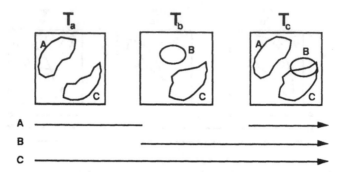

Figure 8.3 Sample area features. Area A dies and is reincarnated. Area B is born, then moves to intersect Area C. Area C changes shape and changes where it intersects Area B.

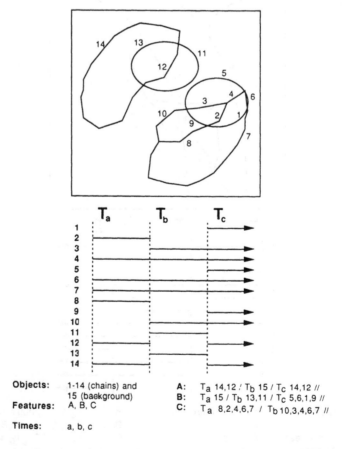

Objects:	1-14 (chains) and	A:	T_a 14,12 / T_b 15 / T_c 14,12 //
	15 (background)	B:	T_a 15 / T_b 13,11 / T_c 5,6,1,9 //
Features:	A, B, C	C:	T_a 8,2,4,6,7 / T_b 10,3,4,6,7 //
Times:	a, b, c		

Figure 8.4 Area features as chains in a space-time composite. A total of 16 chains represents the snapshot sequence of Figure 46. Chains 12 and 14 because they are reincarnated, are counted twice. The 'silver' polygon of Chain 6 is disregarded because it is below tolerance size.

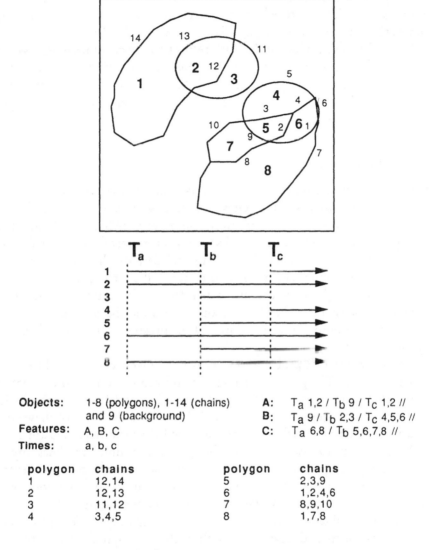

Objects:	1-8 (polygons), 1-14 (chains) and 9 (background)	A:	T_a 1,2 / T_b 9 / T_c 1,2 //
		B:	T_a 9 / T_b 2,3 / T_c 4,5,6 //
Features:	A, B, C	C:	T_a 6,8 / T_b 5,6,7,8 //
Times:	a, b, c		

polygon	chains	polygon	chains
1	12,14	5	2,3,9
2	12,13	6	1,2,4,6
3	11,12	7	8,9,10
4	3,4,5	8	1,7,8

Figure 8.5 Area features as rings in a space-time composite.

greater problems with the access system. These factors become clearer in the implementations that follow. However, the trade-off of not using rings as a processing unit is that response to point-containment queries[26] requires more computation or relies more heavily on the data structure. If rings are stored, response to such a query involves first examining the minimum bounding rectangles of the rings to find which rectangles contain the point, then

---

[26]A point-containment query asks, 'what objects contain this point'?

refining the search to see whether the point lies within the boundary of the ring itself. Without stored rings, the procedure must rely on topological navigation or tree traversal (e.g. Sarnak and Tarjan, 1986; Van Oosterom, 1989) to determine where it lies with respect to the chain network. Whether the benefits of compactness offset added computation depends on the application and other implementation details. Since I am the computer in my implementations, my choice is clear: I can forego assistance in point-containment queries.

What to access?

Chapter 7's discussion on access patterns suggests the need for a scheme to improve access to past data states. To review, access occurs in dual data spaces, one which contains attributes, the other which contains spatial data. The problem of accessing temporal attributes is relatively approachable, since it is addressed by the voluminous temporal database literature reviewed in Chapter 5. Attribute data access also is more likely to be a subset of spatial data access than vice versa because spatial data are dimensional while attribute data are dimensionless. For this reason, I set aside the problem of accessing temporal attribute data to address the greater challenge of accessing spatiotemporal data.

Ironically, the first problem to solve is not what access method to use, but what to access. The goal is to recognize and retrieve only those objects that are needed in a requested time slice or span without referencing the attribute data space. Two questions arise: what 'objects' are the focus of these efforts, and what determines whether they are needed at the requested time?

My decision to use chains as the processing unit simplifies my task considerably. To prepare the atemporal topological model for spatiotemporal indexing requires that chains alone be made temporal because polygons are implicit in the chain representation. An atemporal chain record describes the nodes that terminate the chain and the areas it separates. Adding timestamps to describe birth and death makes that chain temporal. Thus, the chain descriptions are:

Atemporal: ID; from/to node; left/right polygon.
Temporal: ID, from/to node; left/right polygon; birth, death.

An earlier definition of objects hampers this description of a temporal chain. Chapters 4 and 6 define object lifespans as infinite unless a system partitions temporal data into epochs, in which case objects are born and die only at epoch boundaries. This construct seems to preclude the description of a temporal chain offered above. If objects persist indefinitely, how can a chain have timestamps for birth and death? Formally, it cannot; but we can

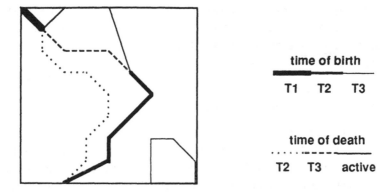

Figure 8.6 The virtual lifespans of chains, as used in the Figure 3.6 example of the space-time composite. Lifespans are determined by when chains are an active part of the representation. A chain's record states the times of virtual birth and death.

circumvent this logical inconsistency by defining virtual lifespans that define object activation periods.

Chapter 3 described how to access a snapshot: obtain the attribute histories of all objects in that area at all times, find the attribute sets in effect for the requested time, then 'dissolve' the unnecessary chains that divide areas of like attributes. Chains that remain after this procedure are the activated objects for that time. Thus, a virtual lifespan is the activation period when an object would not be dissolved by the time-slicing procedure described above. The virtual births and deaths of objects in a feature-based representation correspond to the actual births and deaths of features. In an object-based representation, objects are activated only when their attribute is not 'background' or (in the case of some chains) when they separate two areas with different attributes at that time slice (Figure 8.6).

Ground rules for implementation

Having defined the 'targets' of the access method, all that remains is to define the operational details of access procedures. The test data describe features as the aggregates of objects. In my implementation, objects are rendered by minimum bounding cubes (MBC) that completely but minimally enclose them, and whose x, y, and t dimensions are orthogonal to the x, y, and t axes. The t dimension of the MBC corresponds to an object's virtual lifespan. The MBCs reside in a quasi-physical data space of dimensions x, y, and t.

The basis of partitioning is that each cell in the quasi-physical data space corresponds to a data bucket in storage. Partitions are refined when buckets overflow. In a genuine implementation, bucket capacity would be fixed by a system manager or database designer based on the needs of the system. Rather than establishing arbitrary record and bucket sizes for this exercise, I observe two rules:

1. all objects have the same storage requirements
2. buckets hold a maximum of four objects.

The methods that I examined all build their partitions based on the data that they house. To begin, the data space is unpartitioned, corresponds to one data bucket, and holds all the data. If the total number of objects exceeds bucket capacity, partitions are inserted until all buckets are at or below capacity. Partitions can be refined along any of the three axes, x, y, or t, which means that refinements can favor time or space, or treat them equally. In sum, the placement of partitions should relate to the dimensional dominance of a given application. Because I wanted to examine how and also how well a structure can adjust itself to different levels of dimensional dominance, I experimented with rules for adding partitions when a method seemed generally promising. Partitioning methods differ; standard rules are difficult to phrase. I describe my experiments in later sections that describe the specific partitioning methods.

Accessing the stored data

Retrieval costs are risky to gauge using a qualitative approach such as mine. I discuss retrieval in terms of what is involved, but I have determined what is involved by 'querying' my implementations. I chose two query types to represent two fundamental spatial operations (Faloutsos *et al.*, 1987):

Point containment: given a point, find all objects that contain it.
Regional intersection: given a region, find all objects that lie within or intersect it.

In both cases, I extend the Faloutsos operations to incorporate temporality:

Temporal point containment: given a point, find all objects that contain it during a given time span.
Temporal region intersection: given a region, find all objects that lie within or intersect it during a given time span.

In the light of previous discussions, point-containment queries are important because they could be the starting point for a temporal query if the feature whose versions are sought was initially identified from a display. Region-intersection queries form a basis for the spatiotemporal queries discussed

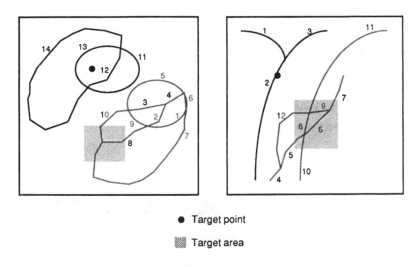

● Target point

▓ Target area

Figure 8.7 Test of retrieval. A point and a region were randomly selected to serve as target objects when evaluating the retrieval efficiency of the access methods implemented here.

previously (for example, to find the information that intersects a space-time cube defined by ranges x, y, and t).

I used two strategies to examine how easily information is retrieved from the data structures I implemented. The first is random, the second is purposeful. The random method was intended to serve mainly as a preliminary exercise, although in some cases no further queries were needed because the results were clearly so poor. I devised the random queries by selecting a point and a region from both the sample line and the sample area data (Figure 8.7). Both random queries target the interval T_a–T_c.

For methods that dispatched the random queries reasonably well, I attempted to consider what the best- and worst-case behaviour would be. To do this, I chose what appeared to be the easiest and the most difficult queries to respond to. An 'easy' query references data stored in the most accessible partition; a 'difficult' query references a region that crosses several inaccessible partitions. In all cases, I estimated the number of buckets accessed and the number of objects examined.

The grid file implementation

I used the grid file to evaluate the effects of a partitioning system that clips objects. The grid file's creators (Nievergelt *et al.*, 1984) did not originally design their method to treat dimensional data, choosing instead to treat

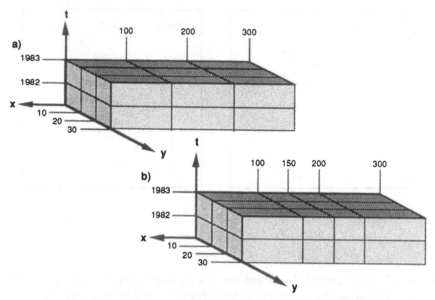

Figure 8.8 Grid files as described by Nievergelt et al. (a) The data space defined by axes x, y, and t is partitioned into cells by appropriate intervals. (b) Partitions along axis x are refined because one or more buckets corresponding to those cells overflowed.

dimensionless data in *k*-dimensional space. I chose the grid file nonethless because it offers strong procedures for treating *k*-dimensional space, has a relatively flexible regular partitioning strategy, is clearly documented, and is elegantly simple.

Mechanics

Grid files are designed to manage storage units—buckets—of fixed size. The grid file partitions data space systematically, and each subdivision is assigned to a bucket (Figure 8.8). Depending on the number of records a cell contains, it may share a bucket with other cells. The reverse is not true, however; one cell can reference no more than one bucket. As explained by Nievergelt *et al.* (1984), the problem reduces to defining the correspondence between grid cells and buckets.

Figure 8.8 shows a three-dimensional data space defined by axes x, y, and z that is partitioned into blocks by intervals a, b, and c. Only if data are distributed uniformly through data space would this mapping of regularly sized cells to buckets be workable. Figure 8.8b shows the same data space with interval $b2$ refined because one or more of the buckets corresponding to its cells overflowed. Further overflows induce further grid refinements.

Several implementation options face a system designer using a grid file.

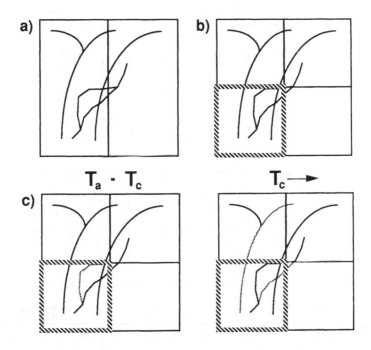

Figure 8.9 First set of grid file partitions. The patterned line indicated bucket overflows. a) the x dimension is partitioned. Four objects are clipped and the two resulting buckets overflow with 11 and 5 objects. b) The y dimension is partitioned. Two more objects are clipped and one bucket still overflows. c) The t dimension is partitioned. Six more objects are clipped and the same two buckets overflow.

The grid file can store the objects themselves or it can reference the objects, which are stored elsewhere. In addition, a grid file can split a clipped object or it can describe the complete object on both sides of the partition that clips it. If the grid file stores the objects themselves, dual representation of clipped objects is inadvisable because of the amount of space involved. The alternative that splits the object requires that the object be reassembled before processing, but this is a far better alternative. Conversely, if the file stores object references, dual representation of the complete object can be more economical because then only one trip to storage is needed to retrieve the entire object. If the object is split and stored, two trips to storage and mending before processing are required. In either case, a clipped object effectively results in a second object.

Precisely how and where to refine grid-file partitions is not strictly prescribed. Nievergelt *et al.* (1984) describe a system of successive refinement that halves any cell that overflows, alternating directions repeatedly until all buckets are at or below capacity (as shown in Figure 7.11). This tactic becomes less effective as dimensions are added because each new partition

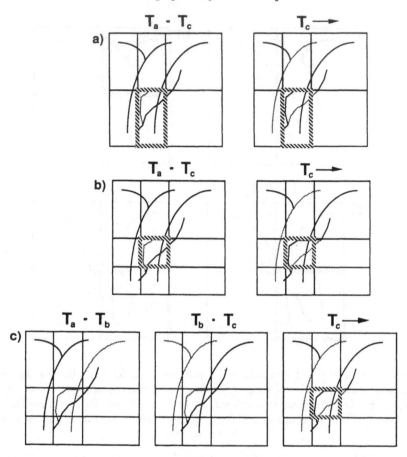

Figure 8.10 Second set of grid file partitions. The patterned line indicates bucket overflow. a) The x dimension is partitioned. Five objects are clipped and the same two buckets overflow. b) The y dimension is partitioned. Five objects are clipped and the overflow is unresolved. c) The t dimension is partitioned. Seven objects are clipped and one bucket still overflows.

increases the total number of cells by the number of cells it crosses, and more cells are crossed as more dimensions are added. Thus, cells that are at capacity are subdivided to below capacity to ease overcrowding in other cells that lie along the same axis as the partition.

Demonstration

Using the sample line data, I imposed a set of spatiotemporal partions. The refinement strategy was to insert new partitions in order x, y, and t until no buckets overflow. The first two partitions cross the two spatial dimensions and leave one cell overflowing (Figure 8.9a and 8.9b). I add a temporal partition, which clips more objects but does not alleviate the overflowing cell (Figure 8.9c). The total number of objects at this stage is 25, for a near doubling of the original 13.

The next series of partitions also cross first space, then time (Figure 8.10). The size of the area that corresponds to the problem bucket decreases, but the overflow persists because object size also is decreasing due to clipping. Now the number of objects has swelled to 42, more than triple the original 13. One bucket is still overflowing and a total of 19 buckets are occupied.

Comments

Implementing this approach on the test data produced an eloquent case against permitting objects to be clipped. The added third dimension rendered a marginal two-dimensional strategy unworkable.[27] Because of spiralling problems caused by clipped objects, I implemented this approach using only the line data and did not bother to 'query' the implementation because it is clearly not suited to treat spatiotemporal data.

The effect of clipped objects is sobering. The magnitude of increased data volume was unexpected. The number of chains needed to represent the line data more than tripled and bucket overflow persisted. Tripling of the data volume is not a freak occurrence. A second implementation (not shown) trisected rather than bisected cells, and nearly tripled the number of objects (to 38 from 13) before finally eliminating bucket overflow. Six and Widmayer (1988) had similar results when they compared the grid file to their offset grid file. Using two data dimensions, they found that the number of objects swelled by factors of approximately 1.5 to 2.5 due to clipping. The problem of clipped objects goes beyond that of volume alone; Kleiner (1989), who compared clipping and non-clipping partitioning methods, found that the effort of reassembling clipped objects was far greater than the time required to retrieve them.

The fundamental inefficiency of clipped objects is unavoidable. However, it is possible to mitigate their effect by adjusting bucket size. My relatively small buckets, designed to illustrate the dynamics of partition refinement, do aggravate the problem. Larger cells, which are possible with larger buckets, clip fewer objects and reduce the spiralling growth factor. Six and Widmayer found that inserting 1211 rectangles into a two-dimensional grid file produced 3080 rectangles when a bucket held 16 rectangles but only 1771 when a bucket held 64.

The refinement strategy also contributed to problems. The Nievergelt

---

[27]The grid file is marginal for dimensional objects. It was designed for dimensionless objects, and for those it is effective.

strategy is too sweeping. Each new partition crosses the entire length of the dimension it divides, even if only a portion of the data space is occupied by an overflowing cell. Added to these sweeping slices is the problem of more dimensions; a partition in one dimension cuts across the partitions of other dimensions, as well. Thus, a new partition subdivides sparsely populated cells along with the overflowing ones that it is intended to relieve. Apparently, each added dimension will worsen this problem and it may be universal to trie approaches.

The effect of the third dimension was overwhelmingly negative. Aside from a multiplier effect with each new partition, problems associated with clipped objects appear to increase with the number of dimensions. The virtual lifespans of objects cross temporal partitions just as their geometric outlines cross the spatial ones. This means that long-lived objects are represented in every temporal partition, which causes buckets to overflow, partitions to refine, and more objects to be clipped.

The offset grid file implementation

Using separate sets of partitions that are systematically offset to avoid clipping objects is one way to avoid the problems described above. I selected the offset grid file of Six and Widmayer (1988) as the vehicle to examine this strategy. The refinement scheme that they recommend also makes more conservative cuts in the data space so overpopulated cells do not cause underpopulation among cells along the same axis. Thus, while this, too, is a trie approach, it avoids the more dramatic problems experienced with the classic grid file.

Mechanics

The Six and Widmayer (1988) enhanced grid file avoids splitting objects because when a partition clips an object, that object is removed to a second file level with different partitions that (hopefully) do not clip it (Figure 8.11). If the partitions of the first and second file levels both clip the object, a third, and even a fourth level can be established with different partitions. By permitting the fourth file level to clip objects, no more than four levels are ever needed.

Six and Widmayer argue that only very large objects will percolate through to the fourth file. Their performance statistics back their arguments and suggest that this enhancement improves considerably upon the standard grid file (as implemented above). For best results, the partitioning strategy for these tiered grid files is critical. The method chosen by Six and

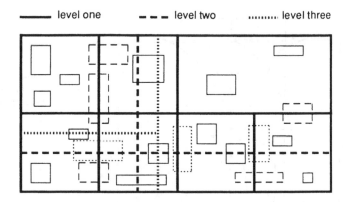

Figure 8.11 Partitions of a three-level offset grid file, where each level's partitions are offset from the next so objects that cross one set of partitions can be stored without splitting at another level.

Widmayer situates each partition of each successive layer to maximize its distance from those of all preceding layers. This produces a hierarchical but unnested set of partitions; the first file level has the finest partitions, the last file level has the coarsest.

Demonstration

Maintaining multiple offset grid files is surprisingly effective in overcoming the clipping problems of the simple grid file. Using a spatiotemporal approach that favors neither space nor time, I partitioned the sample line data into three levels, which resulted in only seven occupied buckets (Figure 8.12). Partitions are refined progressively in the cells that exceed bucket capacity, which for this exercise was arbitrarily set at four chains (Figure 8.13). To treat the three dimensions equitably, refinements occur in sequence: first to the x dimension, next to the y dimension, and finally to the t dimension.

The spatiotemporal implementation's equal treatment of space and time resulted in 'clusters' with no apparent spatial or temporal coherence. Because spatiotemporal clusters are difficult to visualize, and thinking that perhaps the added time dimension was the culprit, I tried a space-dominant partitioning scheme that inserts new partitions along the x and y axes only (Figure 8.14). Figure 8.15 shows the progression of refinements needed to produce Figure 8.14's partitions.

It would appear that this method is incapable of clustering neighboring values. Figure 8.16 shows the buckets that form from spatiotemporal and space-dominant partitioning schemes. In neither case is there any clustering

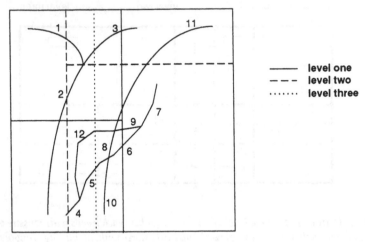

Figure 8.12.

Level One progression

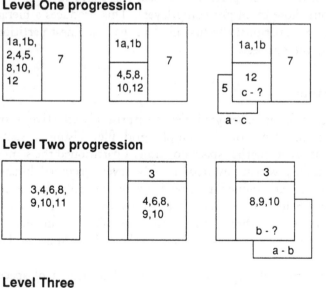

Level Two progression

Level Three

Figure 8.13 Progression of refinement for Figure 8.12's partitions. A '?' indicates the present.

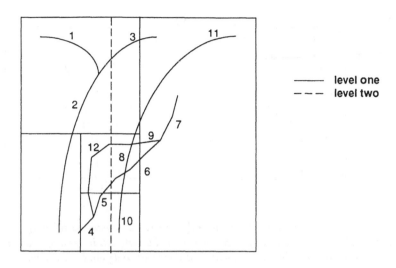

Figure 8.14 Line data in space-dominant offset partitions.

Level One progression

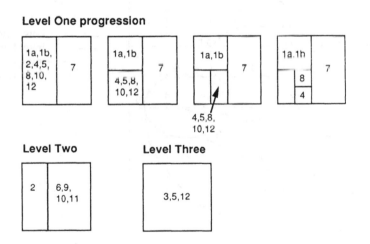

Figure 8.15 Progression of refinement for Figure 8.14.

of neighbors. In fact, despite Six and Widmayer's claim to the contrary, the larger and longer-lived objects do not migrate to the higher file levels.

Retrieving data from the offset grid file exposes the method as a slow-but-steady 'tortoise' (compared to a more erratic 'hare'). 'Random', 'easy', and 'difficult' queries produce virtually the same results: all file levels must be examined. Thus, a point-containment query will always access one bucket per level unless the appropriate cell of a level is empty. A region-intersection query is less predictable because the extent of the region relative to the

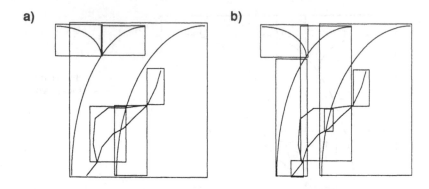

Figure 8.16 Buckets formed by spatiotemporal and space-dominant partitioning schemes. a) Spatiotemporal buckets produced by inserting x, y, and t partitions in turn. b) Space-dominant buckets produced by inserting partitions along the x and y dimensions alone.

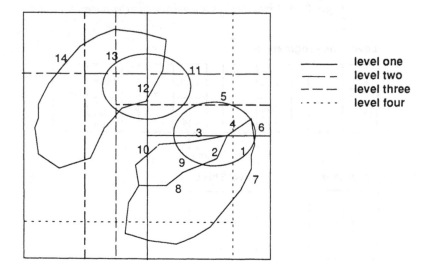

Figure 8.17 Area data in regular offset partitions. Figure 8.18 shows membership of objects to levels and the sequence of partitioning.

placement of partitions determines the buckets to be referenced. The random region query, which seeks all or part of chains 5, 6, 7, 8, 9, 10, and 11, involved six of the seven occupied buckets. Evidently, a 'worst case' query could reference all buckets; then, the fewest total buckets with the highest possible occupancy rate provide the best performance.

The results of the line data experiment were essentials repeated with the area data. The sample areas needed four grid file levels to resolve bucket overflow; the fourth level splits Areas 12 and 14 in the temporal dimension. Figure 8.17 shows the placement of partitions. The sequence of refinements is identical to the spatiotemporal implementation performed for the line data, i.e. the first partition divides the x axis, the second divides the y, and the third divides the t (Figure 8.18). The result of this implementation is that the 16 chains that describe the area features are contained in only seven cells that correspond to seven data buckets.

As with the sample line data, the retrieval performance for the area data is predictably mediocre. I did experience unforeseen problems with the strategy

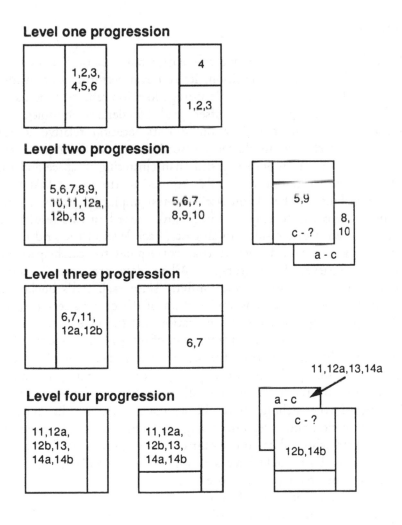

Figure 8.18 Progression of refinement for Figure 8.17.

I adopted for representing areas with chains (not rings). For point-containment queries, if the chains that bound the polygon that contains the point do not lie in the same partition with the point, the search must widen to neighbouring partitions until one of its chains is found to direct further search. The multi-tiered storage strategy of the offset grid file requires that all tiers must first be searched in the partition that houses the point before the search can be widened. If a point were contained in a very large polygon whose chains lay in many different (and distant) partitions, response to such a query could become quite expensive.

Comments

The offset grid file uses storage efficiently; but because it does not cluster data according to relative position, its retrieval performance is mediocre on range searches. Ideally, an object's storage location reflects its actual position in k-dimensional space. The offset grid file defines an object's storage location by both its true location and its location relative to the file's partitions, which gives its clusters a disorderly air. My attempts to cluster in space alone were no more successful than clustering in space and time.

But coherent groupings are only required to the extent that they affect data accessibility. When I 'queried' the test implementations, I found that the data's apparent disarray in storage gives poor best-case results but quite acceptable worst-case results. In all cases, each level of the grid file must be examined and occupied buckets that correspond to search-space locations must be retrieved from storage. When data are sparsely but evenly distributed in cells, performance suffers, since each cell's bucket must be examined if the cell contains any portion of the query's search space.

More positively, there is some indication that the added dimension of time actually speeds access to data stored in an offset grid file. One would suspect that temporality would create the same havoc it caused in the regular grid file, where the problems of clipping increased with the number of dimensions. In the case of the offset partitions, however, the added dimension further delimits both the data space within cells and the search spaces of queries. Since response to a query is, fundamentally, the retrieval of data buckets whose data space intersects the query's search space, adding constraints to both could have a positive effect. While this effect may be offset by more data (hence more buckets), it is a novel idea that more information could improve data accessibility.

While my results are inconclusive, the surprising indication that an added dimension does not necessarily worsen performance merits further investigation.

The R-tree implementation

A final three-dimensional approach is to establish irregular partitions that avoid clipping objects altogether. R-trees (Guttman, 1984) are representative of this strategy, and are used in several operational GISs.

Mechanics

R-trees avoid splitting objects by building a search tree of nested *k*-dimensional rectangles which at the lowest level, are wrapped around the objects themselves (Figure 8.19). The bottom nodes of the search tree reference data buckets, making R-trees a multidimensional extension of the zero-dimensional B-tree (Bayer and McCreight, 1972). The building and balancing processes are governed by heuristics.

R-trees are designed to permit purposeful clustering of dimensional data in multidimensional space. Their greatest asset is their adaptive clustering of data and their general avoidance of clipping objects. R-trees are also intuitively elegant, which minimizes conceptual difficulties in programming. Problems with R-trees relate to their heuristic construction, which can be slow and of varying effectiveness. In addition, R-trees have higher overhead costs than regular partitioning methods because the limits of the irregular partitions must be stored, rather than computed. But the most serious problem of R-trees is the likelihood that their cells will overlap. Overlapping cells mean that small objects can be embedded entirely within more than one cell (e.g. R6 of Figure 8.19), or clipped by a cell partition other than their

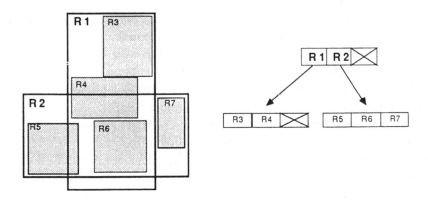

Figure 8.19 The irregular partitions of an R-tree. Rectangle R6 is completely contained and R4 is partly contained by R1 and R2. Queries whose search space includes the overlap area between R and R2 are answered only by examining the children of both subtrees.

own (e.g. R4 of Figure 8.19). Searches for objects of ambiguous ancestry involve following more than one path downward through the tree. In addition, any search space contained within an overlap area must be sought by traversing all parts of the tree that contain it. Overlap is particularly disastrous at the highest tree levels, as in the illustration, because they fail to narrow the search space and induce major loss of efficiency.

Guttman (1984) designed the original R-tree heuristics to minimize total coverage, i.e. the total area of index leaf nodes, but because of the effect of overlap on performance, Roussopoulos and Liefker (1985) developed an algorithm to minimize both coverage and overlap. They demonstrate that it is only possible to eliminate overlap for dimensionless data and only then when data are known in advance, not introduced dynamically. Roussopoulos and Liefker also demonstrate that zero overlap cannot be achieved for dimensional objects. However, their R-trees pack each bucket to capacity and do improve search efficiency. To further address the overlap problem, Faloutsos *et al.* (1987) introduce a partitioning strategy that avoids overlap altogether; ironically, however, it does so only by dual representation of objects that fall in overlap areas. An alternate but related strategy would be to clip objects in overlap areas.

Demonstration

To illustrate the spatiotemporal options, I implemented the line data in R-trees twice: first to minimize overlap in the spatial dimensions (for space-dominant applications) and then in the spatial and temporal dimensions (for spatiotemporal applications). I also implemented the area data to favor neither space nor time.

The first implementation of the line data, which is entirely space-dominant, clusters chains based on spatial proximity and disregards their temporality (Figure 8.20). The line data fit into four buckets with modest spatial overlap. Chains 8 and 9 have ambiguous ancestry because both are situated in overlap areas. Temporal overlap is more severe, however. All four buckets contain data from the complete system lifespan.

The spatiotemporal implementation (Figure 8.21) also requires only four buckets in a tree that is two levels deep (including the root node). Now, however, the overlap is largely spatial rather than temporal. But when all three ranges are used to access the data, none of the chains have ambiguous parentage. Chains 1b and 11, which fall in the temporal range of a second bucket, do not fall in that bucket's spatial range. Likewise Chains 5, 6, 8, 9, and 12, which fall in the spatial range of another bucket but not its temporal range.

The area data were not as amenable to R-tree representation (Figure 8.22).

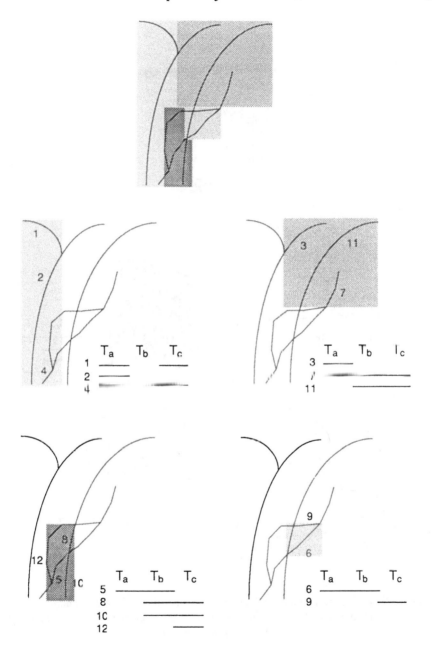

Figure 8.20 The line data in space-dominant irregular partitions. The graphic at the top shows the coverage of each of the four data buckets. The bottom four graphics describe the bucket contents in more detail.

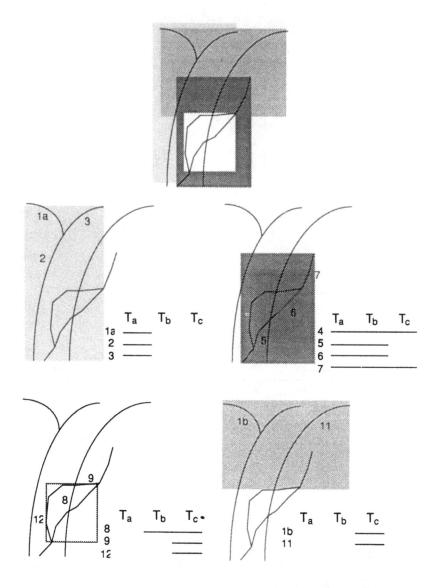

Figure 8.21 Line data in time-dominant irregular partitions. The top graphic shows the coverage of the four data buckets. The bottom four graphics provide detail concerning the contents of those buckets.

The choice is whether to increase coverage and have a shallower tree or to deepen the tree one level and minimize coverage. I chose the latter course. The end result is that the area data's sixteen chains fit into five buckets of a tree three levels deep (including the root node).

Retrieving data from R-tree structures produced more radical best- and

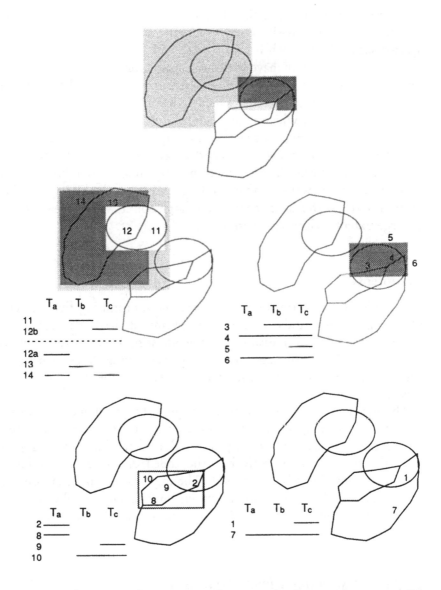

Figure 8.22 Area data in irregular partitions. The top graphic illustrates the coverage of the first level of buckets. One bucket is further subdivided due to overflow.

worst-case performance than the offset grid file, which appears to present a designer with a tortoise-versus-hare dilemma. In the best case, the search space is entirely contained within a single cell that overlaps no other; then, the data are accessed from one bucket. The worst case, however, can involve many overlapping nodes and many bucket retrievals.

Among the more interesting results is an indication that the temporal dimension is not necessarily a negative factor. Adding constraints to the data space associated with each R-tree bucket and on the search space associated with a query seems to have a positive effect on R-tree retrievals and to mitigate overlap problems. In other words, the temporal data added to the spatial actually reduce overlap, which is helpful; however, it also increases the amount of data, which is not helpful.

A side-effect of using the space-time composite to represent temporality is that the decomposition progressively decreases object sizes. Uniformly small objects could decrease the amount of overlap in an R-tree because cells are more focused. However, mixes of large with small objects are likely to worsen R-tree performance because the large objects would create index cells of large extent, which are more likely to intersect.

Comments

The results of the R-tree experiment are more promising than anticipated. My reconnaissance study did not attempt any of the R-tree enhancements that are possible. Such measures include adjusting bucket sizes, packing the R-tree to minimize coverage and overlap, developing different heuristics to define placement of partitions, and representing objects in overlap areas twice.

The added temporal dimension will require that tree-building heuristics be extended to forming cubes of minimal volume (in contrast to the two-dimensional practice of forming rectangles of minimal coverage). These heuristics will be more costly but the added dimension also offers more options in how the cells are formed. The implication that more dimensions do not necessarily imply greater processing difficulty should be explored. Enhancements that exploit this factor could be worth pursuing.

Exploring six-dimensional data space

Having reviewed three options for dealing with dimensional data objects, it is interesting to return to the fourth option that was mentioned earlier, then suspended temporarily. Rather than establishing k_{data} space to equal $k_{objects}$, data space can be defined to have whatever number of dimensions is required to reduce objects to points. Since my strategy uses a temporal chain as the processing unit and the MBC of a temporal chain occupies three dimensions, the required number of data space axes is six: one each for minimum and maximum x, y, and t.

Table 8.1 *Line data stored in regular six space partitions. A 'A' means an open-ended range that includes the present.*

Cell	xmin	xmax	ymin	ymax	tmin	tmax	contents
1	0–12	0–12	0–24	0–24	a–c	a–c	5,2,1a
2	0–12	0–12	0–24	0–24	a–c	c–?	4,8,10
3	0–12	0–12	0–24	0–24	c–?	c–?	12,1b
4	0–12	12–24	0–24	0–24	a–c	a–c	6,3
5	0–12	12–24	0–24	0–24	a–c	c–?	11
6	0–12	12–24	0–24	0–24	c–?	c–?	9
7	12–24	12–24	0–24	0–24	a–c	c–?	7

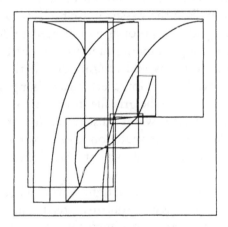

Figure 8.23 *Coverage of six space cells as they would appear in two space.*

Demonstration

If three-dimensional temporal chains are mapped as points in six-dimensional space, any of the methods designed to treat points in k-dimensional space can be applied to the spatiotemporal problem. Table 8.1 maps the temporal lines to a six-dimensional grid file. The cells cover the study area as described in Figure 8.23.

The 13 objects were stored in only seven buckets and the mapping of objects to buckets is fast and simple. While placing dimensional objects into partitions proceeds smoothly in a higher-dimensional space, performing range queries does not. Consider the two test queries. To find the objects that contain a given point requires referencing each cell in six space that has a range that includes any of its three coordinates. The queries that access the line data seek a point in the range of $x = 0\text{-}12$. To find all objects that contain this point requires that each bucket that corresponds to a data space with a minimum or maximum x in that range must be accessed. To respond to the random point-containment query required that five of the seven buckets be retrieved. To respond to the random region-intersection query required that all seven buckets be retrieved.

Comments

It is simple to place dimensionless objects within six-space partitions. An obvious disadvantage—the overhead of three additional data space dimensions—is not as bad as it first appears. MBCs, used by all the schemes, already provide six 'coordinates' for each object, which are readily mapped into six-space. The bulk of the six-space itself relates to how it is treated by a given approach. A grid file approach would need to define its partitions as hyperplanes and store a partitioning function or partition locations for each dimension. These measures would roughly double the size of the grid file's relatively modest overhead. However, maintaining information on six-dimensional cells could cause the grid file directory to grow to unmanageable size. If a hashing scheme were substituted for the directory, the six components would make computation more costly than for the three-dimensional case. However, neither of these expenses appears radically unreasonable.

Most of these comments are moot, however; while storing dimensional data as points presents few problems, retrieving them does. Three-dimensional ranges collapse into dimensionlessness in their journey from three-space to six-space. This means that our focal interests, containment and intersection of spatial objects, become conceptually and technically difficult to treat.

Whether these problems are insurmountable is unclear. A more thorough examination of six-space options could be revealing. Even if retrieval problems were solved, however, the problem remains of how difficult six-dimensional thinking is to most humans. Creating and maintaining software that is designed around six-space would be a considerable risk; unless the pay-offs are also considerable, this approach is inadvisable.

Evaluation

The results of these qualitative experiments included some surprises. One would expect that adding data dimensions would aggravate any problems experienced with fewer dimensions. This expectation proved accurate in the case of the classic grid file because of spiralling problems with clipped objects. However, the other implementations did not suffer from this effect. The offset grid file and the R-tree both seemed to benefit from the constraints that temporal dimensions place on data and search spaces. While this is a preliminary finding, it is also a reasonable one. If proven to be generally true, it may be possible to capitalize on the phenomena when designing or enhancing higher-dimensional data structures.

Poor performers included the grid file and reducing the dimensional objects to dimensionlessness in six-space. The grid file's practice of clipping objects at partitions is untenable when new complexity is added to the data. Clipped objects increase the amount of storage required, which causes buckets to overflow. To resolve this overflow, partitions are refined to decrease cell size. The new partitions clip still more objects, which expands the amount of storage required. This spiralling effect eventually resolves itself, but only after drastic increases in data volume. Arguably, any method that clips objects is inadvisable for local partitioning schemes in general. In contrast to the grid file, the six-space method made the mapping of objects to storage locations nearly effortless. But extracting those objects from storage again is quite challenging because object dimensionality disappears in the higher-dimensional space. This means that a query whose search space is a two-dimensional range in three-space seeks a zero-dimensional 'range' in six-space, and that range could intersect the data space of many different buckets. While it is difficult to abandon an approach so novel, abandonment is necessary because no obvious solutions exist to the retrieval problem.

The two acceptable performers each demand further investigation before satisfactory implementation could be assured. Both the offset grid file and R-tree react well to the third data dimension. Each adds a small amount of overhead to describe planer (not linear) partitions and volumetric (not planer) cells. Preprocessing costs do become a consideration as dimensions increase. Both the offset grid file and the R-tree determine the placement of partitions heuristically. The R-tree's heuristics revolve around the relative locations of the objects and attempt to group them economically. Conversely, the offset grid file's heuristics revolve around the placement of existing partitions and are considerably less complex, but the desired effect can be elusive. In the latter case, a new partition can be added that has no effect on the overflow problem it is intended to solve; Figure 8.9 demonstrates this problem using the Nievergelt partitioning strategy, but it is equally possible using the Six and Widermayer method.

As noted above, the indication that higher data dimensions do not necessarily worsen data-processing problems is notable. We can assume that spatiotemporal data will be denser than spatial data if one is a superset of the other. Higher data density means higher density of partitions if bucket sizes remain constant. We can also expect the total coverage of cells to increase with number of dimensions, a factor that, realistically, should have negative effects. However, each cell now covers a data space of distinct x, y, and t. And each query involves a search space of distinct x, y, and t. For a bucket to be retrieved, its data space must intersect the requested search space, something that becomes less likely as the constraints on each increase. Partitioning heuristics that attempt to create highly distinctive clusters could be fruitful.

A second side-effect of added dimensions is that trie-oriented methods are quite sensitive to their partitioning rules. Sweeping partitions that cut along the length of an entire axis of data space (for example, the x dimensions) will subdivide the cells they cross in all other dimensions (for example, y and t). Thus, many underpopulated cells can be created in the course of subdividing overpopulated cells, with subsequent loss of storage and retrieval efficiency. The most conservative possible cuts will produce the best results, and the benefits of a conservative approach are likely to increase with the number of dimensions.

The result of this exercise is that two reasonable partitioning options emerge for organizing three-dimensional objects. A final issue is how a fully spatiotemporal access method compares to the exhaustive method of collecting all objects in the area then dissolving those not needed. Using an exhaustive approach, a spatial access method would suffice to collect candidate objects, which would be screened for relevance to the requested time span. Chains need not be temporal, although that is a possible enhancement to avoid the 'dissolve' step.

My goal was to explore general approaches by examining specific representatives of each. A legitimate concern is the extent to which my results can be generalized. The poor performers evinced problems that are general to their classes. Clipping is clearly to be avoided. Any method that clips will suffer some inherent inefficiency. At the very least, those who contemplate adopting an access scheme that clips objects should invest in some preliminary metrics so that they proceed with an awareness of how serious is the problem. Six-space methods all share the inherent loss of ranges that I experienced.

Of the two methods with acceptable performance, can we expect to find the same performance characteristics in other members of their class? It is reasonable to assume that regular unnested units (such as the offset grid file) will all lack spatial clustering and require that each level of the hierarchy be accessed. Frank (1983), in describing his regular unnested field tree, notes this problem and delimits search by matching maximum storage levels to thematic characteristics of the data. Kleiner (1989), too, notes the effects on the field tree of clustering by object size rather than by location alone. Thus, it does appear that we can ascribe a tortoise-like response—respectable but not fleet—from all existing regular unnested approaches. Likewise, irregular units (such as the R-tree) must always compensate somehow for the ambiguities of overlap.

The next step

This analysis examined four options for accessing multidimensional data. Two options—irregular partitions and regular but offset partitions—appear to be reasonable approaches to spatiotemporal data access. Two other options—regular partitions that clip objects and mapping data as points in higher-dimensional space—have serious problems without obvious solutions.

My efforts should be considered but a preliminary screening. The phenomenological methodology permitted me to confront the necessary decisions and understand the ingredients of the algorithms involved but did not permit me to gather meaningful performance statistics. To build upon any one of these strategies or to continue to compare them, a more quantitative evaluation would be helpful. Among the components of such an evaluation should be an assessment of the effect of bucket size on storage and retrieval, and some experimentation with partitioning rules or heuristics within the selected frameworks. Of more theoretical interest is the effect of adding further data dimensions, and the comparison of spatiotemporal access methods to exhaustive approaches. Both of the latter topics are rich with research possibilities.

9

Summary and conclusions

This work provides a conceptual, logical, and physical basis for the design of a temporal GIS. The concepts and constructs presented here were designed to be generally applicable to a wide range of systems and applications. When a single general solution was not possible, alternate approaches were presented in terms of trade-offs and selection criteria.

Summary

The initial discussions of Chapters 1 and 2 are conceptual treatments of temporal geographic information, including its form, processing, and potential application. I define a temporal GIS using a framework developed by Sinton (1978). This framework states that representation of the three components of geographic information—location, attribute, and time—is constrained so one component must be fixed. A temporal GIS potentially provides a vehicle for representing geographic information without fixing any of its components. Because a great deal of existing work concerning spatiotemporal subjects fits within Sinton's framework, it falls short of being truly spatiotemporal.

I define the construct of cartographic time in Chapter 3. Cartographic time is a distillation of reality, as cartographic space distills its real-world counterpart. In brief, cartographic time is Newtonian: space and time do not interact. The topology is a set of parallel lines that represent objects as they pass through their versions. Boundaries are measured discretely, even when change is gradual. New terminology describes operands and operators. Cartographic time has three facets that clock events in the world, the database, and on the display, and events can be clocked at all levels of temporal resolution.

I examine common models of spatiotemporality and consider their value for computer representation. I then select and develop one model, and describe its implementation on various types of cartographic data in Chapter 4. The model I select is the spatiotemporal counterpart of the topological

159

model employed in most vector-based GISs today. The model, called a space-time composite, represents change over time as a composite of differences between time slices. A space-time composite consists of a set of greatest common spatiotemporal units tied to attribute histories. Each unit has a history distinct from its neighbors and its geometry is uniform over time. The merits of this approach include that it describes temporal data non-redundantly and it accounts for all space at all times, so errors in the data are detectable. It also does not depart radically from current GIS methods, which should facilitate the upgrade of atemporal systems to temporal ones.

An extensive literature on aspatial temporal databases does exist. Chapter 5 reviews this literature and considers its usefulness to cartographic applications. Much of this work is quite germane because it extends the relational model that is widely used in today's GIS designs. This body of work also addresses temporal logic and its practical application in error control and query languages.

Chapter 6 returns to specifically spatiotemporal design issues. Implementation of a temporal GIS will require institutionalized measures to control data volume and error. The dimensionality of the data involved also poses difficult decisions concerning how to cluster it in storage so that data requests are dispatched promptly. Conceptual discussions on design issues apply to most temporal GIS applications. However, as the discussion turns to implementation it becomes harder to recommend specific measures because they are so dependent on usage and preference. For this reason, I discuss implementation options in terms of their associated costs and benefits, and leave the decision of whether to adopt a given measure to individual system designers and managers.

The space-time composite approach to representation that I support provides quite efficient storage of spatiotemporal information, but as decomposition increases, retrieving snapshots from the database involves progressively more computation. Chapters 7 and 8 investigate methods of accessing the space-time composite that identify which objects are needed at a given time slice without requiring reference to attribute data. An impressive number of purportedly multidimensional access methods exist. While all such methods can operate in multidimensional data space, most cannot operate upon dimensional objects in that data space. Of those that can, two factors are critical to their performance: how they cluster data in storage, and how they search for data when requested. I develop a taxonomy of access schemes based on these two components, which I refer to as 'partitioning' and 'indexing', respectively.

I identify four fundamental indexing methods. Indexes are judged according to the storage overhead they entail and the number of trips to storage required to produce a requested data record. Many acceptable indexing approaches exist in theory. However, actual performance is closely tied to partitioning. For that reason, most of my efforts focus on problems with partitioning spatial and spatiotemporal data.

Only hierarchical partitioning methods can hope to avoid exhaustive searches for data at some level, which are costly and slow. For that reason, I do not consider single-level partitioning schemes as candidates. Of the hierarchical methods, partitions can be regular or irregular. Regular partitions either clip objects or adopt a strategy to avoid clipping. Irregular partitions do not necessarily divide the data space into unique subdivisions. This means that data must be sought in all partitions that cover their portion of data space unless a method provides an alternate search strategy.

I chose four contrasting hierarchical approaches to evaluate for spatiotemporal suitability: regular partitions that clip objects, regular partitions that do not clip objects, irregular partitions, and a scheme that avoids clipping objects by projecting them as points in six-dimensional space. My evaluation was qualitative; I played the role of computer. I structured test data using each of the schemes, then queried each structure. Chapter 8 documents my results.

My findings included some surprises, although they must be considered quite preliminary. Not surprisingly, clipped objects are even more troublesome with three dimensions than with two. The number of objects nearly tripled in one instance because of a spiralling situation of clipping, bucket overflow, and partition refinement, leading to more clipping. Also not surprisingly, the six-space strategy is quite elegant for storing three-dimensional data, but retrieving them is another matter. Because the structure describes the data as dimensionless, intersections and ranges in three or fewer dimensions are virtually unrecognizable. This prevents rapid response to queries that are common to geographic applications.

The other two approaches produced reasonable results. The regular nonclipping partitions maintain acceptable average performance through poor best-case and good worst-case performance. It did appear, however, that the added third dimension may worsen average performance because it tends to push data deeper into the structure. Several measures are possible to improve this method, but it seems destined to be a trustworthy yet unimpressive treatment. The pleasant surprises occurred in the implementation of irregular partitions (which used the R-tree approach). The most serious problem of irregular partitions are that they are not necessarily exclusive of one another, which leads to highly inefficient searches for objects in these overlapping subdivisions of data space. One would expect an added dimension to worsen the problem of overlap. Overlap in one or two of the three dimensions does increase measurably; but if all three dimensions are considered, overlap seems to diminish because of the added constraint on the data space covered by each. This is quite encouraging, although it is also quite preliminary.

The logical continuation of this work

Each aspect of this work—conceptual, logical, and physical—produced a new set of questions that beg to be answered. Analytical problems include how to treat multi-scaled temporal data, how to interpolate and generalize between spatiotemporal samples, what types of analysis could reveal patterns in space and time, and how the past can serve in forecasting the future.

Operational and institutional questions also persist. What role should historical geographic data play in modern government organizations? How long should these data be kept, and by whom? Should the public have access to temporal data, and if so, how? Can an automated system of disseminating incremental changes to data users be implemented, and if so, who would benefit? Can the public contribute incremental updates to assist in temporal data capture?

The temporal GIS framework sketched here lies both within and beyond current theory and practice. If implementation of these concepts is to be widespread, each aspect of the National Standard (Morrison, 1987) must be extended to address specifically temporal concerns. At a basic level, cartographic objects would need temporal and atemporal versions; new terms would be required to describe temporal states, behaviors, and functions; data transfer procedures would require versioning methods and fields for temporal descriptors; quality standards would need to incorporate reporting of temporal accuracy, logical consistency, and completeness, and individual applications would need to consider the essential elements of their features that, if changed, mean death and replacement.

The focus of this work was on digital representation of spatiotemporal information. It did not, however, address graphic representation beyond commenting on how little systematic work has been done. Without the graphic, static or animated, the full benefits of a spatiotemporal system cannot be realized. Ideally, to shield users from the complexity of a system and permit them to focus on their application, a system should be capable of selecting a display format based on the information requested. Automated spatiotemporal display would require that query types be somehow matched to display types by system software.

This work by no means exhausted the topic of digital representation. An interesting conceptual problem is the possibility and potential benefit of representing the full three-dimensional topology of a space-time cube. Other challenges are to adopt a less restrictive definition of cartographic time, for example, one that can treat gradual change. Implementational problems include how to make epochs transparent to the user, and devising seamless procedures for probing the past. Current and potential spatiotemporal operations are another interesting arena of investigation. Today's GISs use a relatively limited set of spatial operators. One might ask how a 'buffering' procedure could exploit the time dimension.

To advance the investigation of direct spatiotemporal access would require a quantitative study to confirm or deny the tentative findings of my qualitative study. The offset grid file or R-tree should be tested by computer implementation using realistically sized datasets before embarking on any development. Of particular use in such a study would be to experiment with the rules or heuristics used to refine partitions. Once satisfactory algorithms are established, the direct access scheme should be compared to a dataflow method of composing time slices. It may well be that each excels for different purposes. A second unresolved issue is the relative benefit of choosing a tortoise versus a hare as a data-structuring methodology. When is consistently mediocre performance preferable to a faster engine that can also falter miserably?

A last topic of interest is the effect of adding dimensions to a given k-dimensional data structure. Do general rules apply to classes of structures? Are the effects on performance consistent with each new dimension? Producing performance statistics to describe these effects would be provocative and could lead to enhancements of k-dimensional methods because of the improved understanding.

A capability to treat temporal spatial information would release cartography from the confines of the two-dimensional page in which it has been trapped over the years. Two parallel tracks would seem to be indicated for advancing such a capability. The first is to introduce rudimentary methods for practical use as soon as possible so that our society can begin to capture a longitudinal view of its geographical behavior. The second is to explore more ambitious and expansive approaches to representing, manipulating, and understanding the evolutionary nature of our world.

References

Aaronson, B. S., 1972, Time, time stance, and existence. In *The Study of Time*, 1, (New York: Springer-Verlag).

Abbott, E. [1884] 1952, *Flatland: a Romance of Many Dimensions*, Sixth edition, (New York: Dover Press).

Abel, D. J., 1986, Bit-interleaved keys as the basis for spatial access in a front-end spatial database management system. In *Proceedings of Auto Carto London*, Volume 1, (Amherst, NY: IGU), pp. 163–177.

Abel, D.J. and Smith, J.L., 1986, A relational GIS database accommodating independent partitionings of the region. In *Proceedings of the Second International Symposium on Spatial Data Handling*, (Amherst, NY: IGU), pp. 213–224.

Abida, M.E. and Lindsay, B.G., 1980, Database snapshots. In *Proceeedings of the Sixth International Conference on Very Large Data Bases*, (New York: IEEE), pp. 86–91.

Adelson-Velskii, G.M. and Landis, E.M., 1962, An algorithm for the organization of information. *Soviet Mathematical Dokl.*, **3**, pp. 1259–1262.

Afsarmanesh, H., McLeod, D., Knapp, D. and Parker, A., 1985, An extensible object-oriented approach to databases for VLSI/CAD. In *Proceedings of the 11th International Conference on Very Large Data Bases*, (Stockholm, IEEE), pp. 13–24.

Ahn, I., 1986, Towards an implementation of database management systems with temporal support. In *Proceedings of the International Conference on Data Engineering*, (New York: IEEE), pp. 374–381.

Ahn, I. and Snodgrass, R., 1986, Performance evaluation of a temporal database management system. In *Proceedings of the SIGMOD '86 Conference*, (New York: ACM), pp. 96–107.

Allen, J.F., 1983, Maintaining knowledge about temporal intervals. *Communications of the ACM*, **26**, pp. 832–843.

Allen, J.F., 1984, Toward a general theory of action and time. *Artificial Intelligence*, **23** (2), pp. 123–154.

Anderson, T.L., 1982, Modeling time at the conceptual level. In *Improving Usability and Responsiveness*, edited by P. Scheuermann, (Jerusalem, Israel: Academic Press), pp. 273–297.

Ariav, G., 1986, A temporally oriented data model. *ACM Transactions on Database Systems*, **11** (4), pp. 499–527.

Armstrong, A.C., 1986, On the fractal dimension of some transient soil properties. *Journal of Soil Science*, **37**, pp. 641–651.

Armstrong, M.P., 1988, Temporality in spatial databases. In *Proceedings of GIS/LIS '88*, Volume 2, (Falls Church, VA: ACSM), pp.880–889.

Arnberger, E., 1974, Problems of an international standardization of a means of communication through cartographic symbols. *International Yearbook of Cartography*, **14**, pp. 19–35.

Aronson, P. and Morehouse, S., 1983, The ARC/INFO map library: a design for a digital geographic database. In *Proceedings of Auto-Carto Six*, Volume 1, (Ottawa: Steering Committee of the Sixth International Symposium on Automated Cartography), pp. 372–382.

Atre, S., 1980, *Data Base: Structured Techniques for Design, Performance, Management*, (New York: John Wiley and Sons).

Ayala, D., Brunet, P., Juan, R. and Navazo, I., 1985, Object representation by means of nonminimal division quadtrees and oct-trees. *ACM Transactions on Graphics*, **4**, pp. 41–59.

Ballard, D.H., 1981, Strip trees: a hierarchical representation for curves. *Communications of the ACM*, **24**, pp. 310–321.

Barbic, F. and Pernici, B., 1985, Time modeling in office information systems. In *Proceedings of the SIGMOD '85 Conference*, (New York: ACM), pp. 51–62.

Basoglu, U. and Morrison, J., 1978, The efficient hierarchical data structure for the U.S. historical boundary file. In *Harvard Papers on GIS*, Volume 4, edited by G. Dutton, (Reading, Massachusetts: Addison-Wesley).

Bauer, K. W., 1984, Public planning and engineering: the role of maps and the geodetic base. In *Modernizing Land Information Systems in North America*. Institute of Environmental Studies Report 123, University of Wisconsin, Madison, pp. 130–139.

Bayer, R. and McCreight, E. M., 1972, Organization and maintenance of large ordered indices. *Acta Informatica*, **1** (3), pp. 173–189.

Beard, M. K., 1987, How to survive on a single detailed database. In *Proceedings of Auto-Carto 8*, (Falls Church, VA: ACSM), pp. 211–220.

Beard, M. K., 1989, Use error? the neglected error component. In *Proceedings of Auto-Carto 9* (Falls Church, VA: ACSM), pp. 808–817.

Beard, M. K. and Chrisman, N. R., 1988, Zipper: a localized approach to edgematching. *The American Cartographer*, **15**, pp. 163–172.

Beech, D. and Mahbod, B., 1988, Generalized version control in an object-oriented database. In *Proceedings of the Fourth Conference on Data Engineering*, Los Angeles, (New York: IEEE), pp. 14–22.

Ben-Ari, M., Manna, Z., and Phueli, A., 1981, The temporal logic of branching time. In *Proceedings of the Eighth ACM POPL Symposium*, (New York: ACM), pp. 164–176.

Bennett, R. J., 1979, *Spatial Time Series* (London: Pion Ltd).

Bentley, J. L., 1975, Multidimensional binary search trees used for associative searching. *Communications of the ACM*, **18**, pp. 509–517.

Bentley, J. L., 1980, Multidimensional divide-and-conquer. *Communications of the ACM*, **23**, pp. 214–228.

Bentley, J. L. and Friedman, J. H., 1979, Data structures for range searching. *Computing Surveys*, **11** (4), pp. 397–409.

Ben Zvi, J., 1982, *The time relational model*. PhD dissertation, University of California.

Berry, B., 1964, Approaches to regional analysis: a synthesis. *Annals of the Association of American Geographers*, **54**, pp. 2–11.

Berry, B. J. L., 1974, Short-term housing cycles in a dualistic metropolis. In *Contemporary Urbanization*, edited by G. Gappert and R. Rose, (California: Sage), pp. 165–182.

Board, C., 1974, Cartographic communication and standardization. *International Yearbook of Cartography*, **14**, pp. 229–238.

Bobrow, D. G. and Katz, R. H., 1986, Context structures/versioning: a survey. In *On Knowledge Base Management Systems*, edited by M. L. Brodie and J. Mylopoulos, (New York: Springer-Verlag), pp. 453–461.

Bolour, A. and Dekeyser, L. J., 1983, Abstractions in temporal information. *Information Systems*, **8** (1), pp. 41–49.

Borchert, J. R., 1987, Maps, geography, and geographers. *The Professional Geographer*, **39**, pp. 387–389.

Borodin, A., 1973, Computational complexity: theory and practice. In *Currents in the Theory of Computation*, edited by A. V. Aho, (Englewood Cliffs, New Jersey: Prentice-Hall), pp. 35–81.

Bouillé F., 1978, Structuring cartographic data and spatial processes with the hypergraph-based data structure. In *Harvard Papers on GIS*, Volume 5, edited by G. Dutton, (Reading, Massachusetts: Addison-Wesley).

Box, E. O., 1981, *Macroclimate and Plant Forms: An Introduction to Predictive Modeling in Phytogeography*. (The Hague: Dr. W. Junk).

Brassel, K., 1975, Neighborhood computations for large sets of data points. In *Proceedings of Auto Carto 3*, (Falls Church, VA: ACSM), pp. 337–345.

Brassel, K., 1978, A topological data structure for multi-element map processing. In *Harvard Papers on GIS*, Volume 4, edited by G. Dutton, (Reading, Massachusetts: Addison-Wesley).

Brent, R. P., 1973, Reducing the retrieval time of scatter storage techniques. *Communications of the ACM*, **16**, pp. 105–109.

Brodie, M. L., 1984, On the development of data models. In *On Conceptual Modeling*, edited by M. L. Brodie and J. Mylopoulos, (New York: Springer-Verlag), pp. 19–47.

Bullock, N., Dickens, P., Shapcott, M. and Steadman, P., 1974, Time budgets and models of urban activity patterns. *Social Trends*, **5**, pp. 45–63.

Burrough, P. A., 1981, Fractal dimensions of landscapes and other environmental data. *Nature*, **294**, pp. 240–242.

Burrough, P. A., 1983, Multi-scale sources of spatial variation in soil: 2. a non-brownian fractal model and its application to soil survey. *Journal of Soil Science*, **34**, pp. 599–620.

Burrough, P. A., 1986, Five reasons why geographical information systems are not being used efficiently for land resources assessment. In *Proceedings of Auto Carto London*, Volume 2, (London: Auto Carto London), pp. 139–148.

Burrough, P. A., 1987, Multiple sources of spatial variation and how to deal with them. In *Proceedings of Auto-Carto 8*, (Falls Church, VA: ACSM), pp. 145–154.

Burrough, P. A., Van Deusen, W. and Heuvalink, G., 1988, Linking spatial process models and GIS: a marriage of convenience or a blossoming partnership? In *Proceedings of GIS/LIS '88*, Volume 2, San Antonio (Falls Church, VA: ACSM), pp. 598–607.

Burton, W., 1977, Representation of many-sided polygonal lines for rapid processing. *Communications of the ACM*, **20**, pp. 162–171.

Buttenfield, B. P., 1984, *Line structure in graphic and geographic space*. PhD dissertation, University of Washington, Seattle.

Buttenfield, B. P., 1986, Digital definitions of scale-dependent line structure. In *Proceedings of Auto Carto London*, Volume 1, (London: Auto Carto London), pp. 497–506.

Buttenfield, B. P., 1987, Automating the identification of cartographic lines. *The American Cartographer*, **14**, pp. 7–20.

Calkins, H. W., 1984, Space-time data display techniques. In *Proceedings of the First International Symposium on Spatial Data Handling*, (Amherst, NY: IGU), pp. 324–331.

Calkins, H. W. and Dickinson, H. J., 1987, The effective use of color in cartographic displays. In *Proceedings of the International GIS Symposium*, Volume 3, Virginia, (Washington DC: NASA), pp. 189–199.

Carlson, W. W., 1985, Algorithmic performance of dataflow multiprocessors. *Computer*, **18**, pp. 30–40.

Carmichael, G. R. and Peters, L. K., 1984, Eulerian modeling of the transport and chemical processes affecting long-range transport of sulfur dioxide and sulfate. In *Modeling of Total Acid Precipitation Impacts*, edited by J. L. Schnoor, (Boston: Buttenworth), pp. 151–179.

Castilho, J. M. V., Casanova, M. A. and Furtado, A. L., 1982, A temporal framework for database specifications. In *Proceedings of the Eighth International Conference on Very Large Data Bases*, (New York: IEEE), pp. 280–291.

Cebrian de Miguel, J. A., 1983, Application of a model of dynamic cartography to the study of the evolution of population density in Spain from 1900 to 1981. In *Proceedings of Auto-Carto 6*, Volume 2, (Ottawa: Steering Committee of the Sixth International Symposium on Automated Cartography), pp. 475–483.

Chen, Z. T. and Pequet, D., 1985, Quad tree spatial spectra guide: a fast spatial heuristic search in a large GIS. In *Proceedings of Auto-Carto 7*, (Falls Church, VA: ACSM), pp. 75–82.

Child, J. C., 1984, *Creating a world: the poetics of cartography*. PhD dissertation, University of Washington, Seattle.

Chrisman, N. R., 1974, Impact of data structure on geographic information processing. In *Proceedings of Auto-Carto 1*, (Falls Church, VA: ACSM), pp. 43–52.

Chrisman, N. R., 1979, The role of algorithmic complexity in geographical analysis. (Unpublished paper).

Chrisman, N. R., 1982, *Methods of spatial analysis based on error in categorical maps*. PhD Dissertation, University of Bristol.

Chrisman, N. R., 1983, The role of quality information in the long-term functioning of a geographic information system. In *Proceedings of Auto Carto 6*, Volume 2, (Ottawa: Steering Committee of the Sixth International Symposium on Automated Cartography), pp. 303–321.

Chrisman, N. R., 1984a, Alternatives for specifying quality standards for digital cartographic data. In *Report 4*, edited by H. Moellering, National Committee for Digital Cartographic Data Standards, Columbus, Ohio, pp. 43–71.

Chrisman, R., 1984b, On storage of coordinates in geographic information systems. *Geoprocessing*, 2, pp. 259–270.

Chrisman, N. R., 1990, Deficiencies of sheets and tiles: building sheetless databases. *International Journal of GIS*, 4 (2), pp. 157–167.

Christ, F., 1978, A program for the fully automated displacement of point and line features in cartographic generalization. *Nachrichten Aus dem Karten: und Vermessungswesen*, 2 (35), pp.5–30.

Clampett, H., 1964, Randomized binary searching with tree structures. *Communications of the ACM*, 7, pp. 163–165.

Clarke, K. C., 1986, Advances in geographic information systems. *Computers, Environment and Urban Systems*, 10 (3/4), pp. 175–184.

Cleveland, W. S. and McGill, R., 1986, Graphical perception and graphical methods for analyzing scientific data, *Science*, 229, pp. 828–833.

Cliff, A. D., Haggett, P., Martin, R. L. and Ord, J. K., 1979, *Spatial Diffusion*. (Cambridge: Cambridge University Press).

Cliff, A. D. and Ord, J. K., 1981, *Spatial Processes: Models and Applications*. (London: Pion).

Clifford, J. and Warren, D. S., 1983, Formal semantics for time in databases. *ACM Transactions on Database Systems*, 8 (2), pp. 214–254.

Clifford, J. and Tansel, A. U., 1985, On an algebra for historical relational databases. In *Proceedings of the SIGMOD '85 Conference* (New York: ACM), pp. 247–265.

Clifford, J. and Croker, A. 1987, The historical data model and algebra based on lifespans. In *Proceedings of the International Conference on Data Engineering*, (New York: IEEE), pp. 528–537.

Codd, E. F., 1968, *Cellular Automata*, (New York: Academic Press).

Codd, E. F., 1970, A relational model of data for large shared data banks. *Communications of the ACM*, **13**, pp. 377–387.

Codd, E. F., 1979, Extending the database relational model to capture more meaning. *ACM Transactions on Database Systems*, **4** (4), pp. 397–434.

Codd, E. F., 1981, Data models in database management. *SIGMOD Record*, **11** (2), pp. 112–114.

Cole, J. P. and King, C. A. M., 1968, *Quantitative Geography*. (New York: John Wiley and Sons).

Colwell, R. N., (ed.), 1983, *Manual of Remote Sensing*, second edition. (Falls Church, Virginia: American Society of Photogrammetry and Remote Sensing).

Comer, D., 1979, The ubiquitous B-tree. *ACM Computing Surveys*, **11** (2), pp. 142–159.

Comins, H. N. and Hassell, M. P., 1976, Prey-predator models in spatially heterogenous environments. *Journal of Theoretical Biology*, **48**, pp. 75–83.

Cooke, D. F., 1985, Vehicle navigation appliances. In *Proceedings of Auto-Carto 7*, (Falls Church, VA: ACSM), pp. 108–115.

Cooke, D. F. and Dawes, B. G., 1987, The dbmap system. In *Proceedings of Auto-Carto 8*, (Falls Church, VA: ACSM), pp. 363–369.

Cooke, D. F. and Maxfield, W. F., 1967, The development of a geographic base file and its uses for mapping. In *Proceedings of the Fifth Annual Conference of the Urban and Regional Information Systems Association*, (Washington DC: URISA), pp. 207–218.

Copeland, G., 1982, What if mass storage were free? *Computer*, **15**, pp. 27–35.

Corbett, J. P., 1979, Topological principles in cartography. U.S. Bureau of the Census, Washington D.C.

Corson-Rikert, J., 1987, Updating multi-layer vector databases. In *Proceedings of the International GIS Symposium*, Volume 2, (Falls Church, VA: IGU), pp. 165–174.

Couclelis, H., 1985, Cellular worlds: a framework for modeling micro-macro dynamics. *Environment and Planning A*, **17**, pp. 585–596.

Cox, N. J., Aldred, B. K. and Rhind, D. W., 1980, A relational data base system and a proposal for a geographical data type. *Geo-Processing*, **1**, pp. 217–229.

Crain, I. K. and MacDonald, C. L., 1983, From land inventory to land management: the evolution of an operational GIS. In *Proceedings of Auto-Carto 6*, Volume 2, (Ottawa: Steering Committee of the Sixth International Symposium on Automated Cartography), pp. 41–50.

Dacey, M. F., 1970, Linguistic aspects of maps and geographic information. *Ontario Geography*, **5**, pp. 71–80.

Dadum, P., Lum, V. and Werner, H. D., 1984, Integration of time versions into a relational database system. In *Proceedings of the Tenth International Conference on Very Large Data Bases*, (New York: IEEE), pp. 509–522.

Dangermond, J., 1984, A classification of software components commonly used in geographic information systems. In *Proceedings of the U.S./Australia Workshop on Design and Implementation of Computer-Based Geographic Information Systems*, (Amherst: IGU Commission on Geographical Data Sensing and Processing), pp. 70–91.

Dangermond, J. and Freedman, C., 1984, Findings regarding a conceptual model of a municipal data base and implications for software design. In *Modernizing Land Information Systems in North America*, Institute of Environmental Studies Report 123, University of Wisconsin, Madison, pp. 12–21.

Date, C. J., 1983, *Database: A Primer*. (Reading, Massachusetts: Addison-Wesley).

Davis, W. M., 1899, The geographical cycle. *The Geographical Journal*, **14**, pp. 481–504.

Davis, W. A. and Hang, C. H., 1986, Organization and indexing of spatial data. In *Proceedings of the Second International Symposium on Spatial Data Handling*, Seattle (Amherst, NY: IGU), pp. 5–14.

Dayal, U. and Smith, J. M., 1986, PROBE: a knowledge-oriented database management system. In *On Knowledge Base Management Systems*, edited by M. L. Brodie and J. Mylopoulos, (New York: Springer Verlag), pp. 227–258.

De, S., Pan, S. and Whinston, A. B., 1985, Natural language query processing in a temporal database. *Data and Knowledge Engineering*, **1**, pp. 3–15.

Delisle, N. and Schwartz, M., 1986, Neptune: a hypertext system for CAD applications. In *Proceedings of the SIGMOD '86 Conference*, (New York: ACM), pp. 132–139.

Dennis, J. B., 1980, Data flow supercomputers. *Computer*, **13**, (November), pp. 48–56.

Dobkin, D. and Lipton, R. J., 1976, Multidimensional searching problems. *SIAM Journal of Computing*, **5** (3), pp. 181–186.

Dobson, M., 1983, Human factors in the cartographic design of real-time color displays. In *Proceedings of Auto-Carto Six*, Volume 1, (Ottawa: Steering Committee of the Sixth International Symposium on Automated Cartography), pp. 421–426.

Dougenik, J., 1980, WHIRLPOOL: a geometric processor for polygon coverage data. In *Proceedings of Auto-Carto 4*, Volume 2, (Falls Church, VA: ACSM), pp. 304–311.

Drinnan, C. H., 1985, Mapping information management system design considerations. *Design Graphics World*, **9** (3), pp. 16–21.

Dueker, K., 1979, Land information systems: a review of fifteen years experience. *Geo-processing*, **1**, pp. 105–128.

Dueker, K., 1985, Geographic information systems: toward a geo-relational structure. In *Proceedings of Auto-Carto 7*, Washington D.C. (Falls Church VA: ACSM), pp. 172–177.

Engstrom-Heg, V. L., 1970, Predation, competition, and environmental variables. *Journal of Theoretical Biology*, **27**, pp. 175–195.

Enslin, W., Ton, J. and Jain, A., 1987, Land cover change detection using a GIS-guided, feature based classification of Landsat Thematic Mapper data. In *Proceedings of the ASPRS-ACSM Annual Convention*, Volume 6, (Falls Church, VA: ASPRS-ACSM), pp. 108–120.

Fagin, R., Nievergelt, J., Pippenger, N. and Strong, R., 1979, Extendible hashing—a fast access method for dynamic files. *ACM Transactions on Database Systems*, **4** (3), pp. 315–344.

Faloutsos, C., Sellis, T. and Roussopoulos, N., 1987, Analysis of object-oriented spatial access methods. In *Proceedings of the SIGMOD '87 Conference*, (New York: ACM), pp. 426–439.

Ferg, S., 1985, Modeling the time dimension in an entity-relationship diagram. In *Proceedings of the Fourth International Conference on Entity-Relationship Approach*, (New York: Elsevier), pp. 280–286.

Fifield, J. A., 1987, The Salt River Project's land use model: a multidimensional tool for allocating forecasted land use types. In *Proceedings of the International GIS Symposium*, Volume 3, (Washington DC: NASA), pp. 293–305.

Finkel, R. A. and Bentley, J. L., 1974, Quad trees: a data structure for retrieval on composite keys. *Acta Informatica*, **4**, pp.1–9.

Fleming, D. K., 1976, Imagination of movement. *Maritime Studies Management*, **3**, pp. 193–194.

Floyd, R. W., 1964, Algorithm 245: Treesort 3. *Communications of the ACM*, **7**, p. 701.

Fowler, R. J., 1978, Approaches to multidimensional searching. In *Harvard Papers on GIS*, Volume 4, edited by G. Dutton, (Reading, Massachusetts: Addison-Wesley).

Frank, A., 1983, Storage methods for space-related data: the field tree. Institut fur Geodasie und Photogrammetrie, ETH, Zurich, Bericht NR 71.

Frank, A. U., 1988, Requirements for a database management system for a geographic information system. *Photogrammetric Engineering and Remote Sensing*, **54** (11), pp. 1557–1564.

Freeman, E. and Sellons, W., 1971, *Basic issues in the philosophy of time*. (LaSalle, Illinois: Open Count).

Freeston, M., 1987, The BANG file: a new kind of grid file. In *Proceedings of the SIGMOD '87 Conference*, San Francisco, (New York: ACM), pp. 260–269.

French, R. L., 1987, Automobile navigation in the past, present, and future. In *Proceedings of Auto-Carto 8*, (Falls Church, VA: ACSM), pp. 542–551.

Fries, O., Mehlhorn, K., Naher, S. and Tsakalidis, A., 1987, A log log n data structure for three-sided range queries. *Information Processing Letters*, **25**, pp. 269–273.

Fuchs, H., Kedem, Z. and Naylor, B., 1980, On visible surface generation by a priori tree structures. *Computer Graphics*, **14**, (3), pp. 92–119.

Fuchs, H., Abram, G. D. and Grant, E. D., 1983, Near real-time shaded display of rigid objects. *Computer Graphics*, **17** (3), pp. 65–72.

Fung, T. and LeDrew, E., 1987, Land cover change detection with Landsat MSS and TM data in the Kitchener-Waterloo area, Canada. In *Proceedings of the ASPRS-ACSM Annual Convention*, Volume 6, Baltimore, (Falls Church, VA: ASPRS-ACSM), pp. 81–89.

Gadia, S. K., 1986, Toward a multihomogeneous model for a temporal database. In *Proceedings of the International Conference on Data Engineering*, (New York: IEEE), pp. 390–397.

Gardner, M., 1970, The fantastic combinations of John Conway's new solitaire game 'Life'. *Scientific American*, **223** (4), pp. 120–123.

Gargantini, I., 1982, Linear oct-trees for fast processing of three-dimensional objects. *Computer Graphics and Image Processing*, **20**, pp. 365–374.

Garner, B., 1983, Towards model-oriented GIS for evaluating the socio-economic effects of resource developments. In *Proceedings of the United States/Australia Workshop on Design and Implementation of Computer-Based GIS*, (Amherst, NY: IGU Commission on Geographic Data Sensing and Processing).

Gerber, R., 1981, Competence and performance in cartographic language. *The Cartographic Journal*, **18**, pp. 104–111.

Gleick, J., 1987, *Chaos*. (New York: Viking).

Gold, C. M., 1978, The practical generation and use of geographic triangular element data structures. In *Harvard Papers on GIS*, Volume 5, edited by G. Dutton, (Reading, Massachusetts: Addison-Wesley).

Gonnet, G. H., Rogers, L. D. and George, J. A., 1980, An algorithmic and complexity analaysis of interpolation search. *Acta Informatica*, **13**, (1), pp. 39–52.

Goodchild, M., 1980, Fractals and the accuracy of geographical measures. *Mathematical Geology*, **12**, (2) pp. 85–98.

Goodchild, M., 1982, Accuracy and spatial resolution: critical dimensions for geoprocessing. In *Computer Aided Cartography and Geographic Information Processing: Hope and Realism*, edited by D. Douglas and A. R. Boyle, (Ottawa: University of Ottawa), pp. 87–90.

Goodchild, M., 1989, Optimal tiling for large cartographic databases. in *Proceedings of Auto-Carto 9*, (Falls Church, VA: ACSM), pp. 444–451.

Goodchild, M. F. and Grandfield, A. W., 1983, Optimizing raster storage: an example of four alternatives. In *Proceedings of Auto-Carto Six*, Volume 1, (Ottawa: Steering Committee of the Sixth International Symposium on Automated Cartography), pp. 400–407.

Green, P. J. and Sibson, R., 1979, Computing dirichlet tesselations in the plane. *The Computer Journal*, **21** (2), pp. 168–173.

Grelot, J.-P. and Chambon, P., 1984, Up-dating a land-use inventory. In *Proceedings of Auto Carto London*, Volume 2, (London: Auto Carto London), pp. 44–51.

Guelke, L., 1979, Perception, meaning, and cartographic design. *The Canadian Cartographer*, **16**, pp. 61–69.

Gunther, O., 1986, The cell tree: an index for geometric data. Electronic Research Laboratory, Memorandum Number UCB/ERL M86/89. College of Engineering, University of California.

Guptill, S. (ed.), 1988, *A process for evaluating geographic information systems*. Technology Exchange Working Group, Technical Report 1, Federal Interagency Coordinating Committee on Digital Cartography, U.S. Geological Survey Open-File Report pp. 88–1105.

Guttman, A., 1984, R-trees: a dynamic index structure for spatial searching. In *Proceedings of the SIGMOD '84 Conference*, (New York: ACM), pp. 47–57.

Hagerstrand, T., 1952, The propogation of innovation waves. *Lund Studies in Geography B, Human Geography*, **4**, pp. 3–19.

Hagerstrand, T., 1970, What about people in regional science? *Papers of the Regional Science Association*, **24**, pp. 7–21.

Hagerstrand, T., 1974, On socio-technical ecology and the study of innovations. *Ethnologia Europa*, **7**, pp. 25–37.

Haggett, P., 1975, Simple epidemics in human population. In *Processes in Physical and Human geography*, edited by R. Peel, M. Chisholm and P. Haggett, (London: Heinemann Educational Books), pp. 373–396.

Haggett, P., Cliff, A. D. and Frey, A., 1977, *Locational Models*. (New York: Halstead Press).

Harrison, G. W., 1979, Stability under environmental stress: resistance, resilience, persistence, and variability. *American Naturalist*, **113**, pp. 659–669.

Harvard University, 1976, *IMGRID Reference Manual*. Graduate School of Design, Cambridge, Massachusetts.

Harvey, D., 1969, *Explanation in Geography*. (New York: St. Martin's Press).

Head, C. G., 1984, The map as a natural language: a paradigm for understanding. *Cartographica*, **21** (1), pp. 1–32.

Henderson, L. D., 1983, *The Fourth Dimension and Non-Euclidean Geometry in Modern Art*. (Princeton, New Jersey: Princeton University Press).

Herring, J., 1987, TIGRIS: topologically integrated geographic information system. In *Proceedings of Auto-Carto 8*, (Falls Church, VA: ACSM), pp. 282–291.

Hillis, W. D. and Steele, G. L. Jr., 1986, Data parallel algorithms. *Communications of the ACM*, **29**, pp. 1170–1183.

Hinrichs, K. and Nievergelt, J., 1983, The grid file: a data structure designed to support proximity queries on spatial objects. Report Number 54, Institut fur Informatic, Eidgenossische Technische Hochschule, Zurich.

Hoaglin, D. C., Mosteller, F. and Tukey, J. W., 1983, *Understanding Robust and Exploratory Data Analysis*. (New York: John Wiley and Sons).

Hoare, C. A. R., 1961, Partition, Algorithm 63; Quicksort, Algorithm 64; Find, Algorithm 65. *Communications of the ACM*, **4**, p. 319.

Holloway, D. P., 1988, Land records management in North Carolina: state involvement in GIS specifications. In *Proceedings of GIS/LIS '88*, Volume 2, (Falls Church, VA: ACSM), pp. 922–931.

Hopgood, F. R. A. and Davenport, J., 1972, The quadratic hash method where the table size is a power of 2. *The Computer Journal*, **15** (4), p. 81.

Hoyt, H., 1970, *One hundred years of land values in Chicago: the relationship of the growth of Chicago to the rise in its land values, 1830–1933*. (New York: Arno Press).

Hunter, G. J., 1988, Noncurrent data and GIS: a case for data retention. *International Journal of GIS*, **2**, pp. 281–291.

Hunter, G. M. and Steiglitz, R., 1979, Operations on images using quadtrees. *IEEE Pattern Analysis and Machine Intelligence*, **1**, pp. 145–153.

Isard, W., 1970, On notions and models of time. *Papers of the Regional Science Association*, **25**, pp. 7–32.

Isard, W., 1971, On relativity theory of time-space models. *Papers of the Regional Science Association*, **26**, pp. 7–24.

Jackins, C. L. and Tanimoto, S. L., 1980, Oct-trees and their use in representing three-dimensional objects. *Computer Graphics and Image Processing*, **14**, pp. 249–270.

Jackins, C. L. and Tanimoto, S. L., 1983, Quad-trees, oct-trees, and k-trees—a generalized approach to recursive decomposition of Euclidean space. *IEEE Transactions on Pattern Analysis and Machine Intelligence*, **PAMI-5**, pp. 533–539.

Johnson, J. H., 1982, The logic of speculative discourse: time, prediction, and strategic planning. *Environment and Planning B*, **9**, pp. 269–294.

Johnson, T. R. and Siderelis, K. C., 1989, Establishing a corporate GIS data base from multiple GIS project data sets. In *Proceedings of Auto-Carto 9*, (Falls Church, VA: ACSM) pp. 874–879.

Jones, S., Mason, P. and Stamper, R., 1979, LEGOL 2.0: a relational specification language for complex rules. *Information Systems*, **4** (4), pp. 293–305.

Jones, S. and Mason, P. J., 1980, Handling the time dimension in a database. In *Proceedings of the International Conference on Databases. British Computer Society*, (London: British Computer Society), pp. 65–83.

Kadmon, N., 1972, Automated selection of settlements in map generalization. *The Cartographic Journal*, **9**,, pp. 93–98.

Kahn, K. and Gorry, G. A., 1977, Mechanizing temporal knowledge. *Artificial Intelligence*, **9**, pp. 87–108.

Karlsson, R. G. and Overmars, M. H., 1988, Normalized divide-and-conquer: a scaling technique for solving multi-dimensional problems. *Information Processing Letters*, **26**, pp. 307–312.

Karlton, P. L., Fuller, S. H., Scroggs, R. E. and Kachler, E. B., 1976, Performance of height-balanced trees. *Communications of the ACM*, **19**, pp. 23–28.

Katz, R. H., Chang, Ellis and Bhateja, R., 1986a, Version modeling concepts for computer-aided design databases. In *Proceedings of the SIGMOD '86 Conference*, (New York: ACM), pp. 379–386.

Katz, R. H., Anwarrudin, M. and Chang, E., 1986b, Organizing a design database across time. In *On Knowledge Base Management Systems*, edited by M. L. Brodie and J. Mylopoulos. (New York: Springer-Verlag), pp. 287–296.

Kedem, G., 1982, The quad-CIF tree: a data structure for hierarchical online algorithms. In *Proceedings of the Design Automation Conference*, (New York: ACM), pp. 352–357.

Keegan, H. and Aronson, P., 1985, Considerations in the design and maintenance of a digital geographic library. In *Proceedings of Auto-Carto 7*, Washington D.C., (Falls Church, VA: ACSM), pp. 313–321.

Kennedy-Smith, G. M., 1986, Data quality: a management philosophy. In *Proceedings of Auto Carto London*, Volume I, (London: Auto Carto London), pp. 381–390.

Kent, W., 1978, *Data and Reality*. (New York: North-Holland).

Kent, W., 1982, Choices in practical data design. In *Proceedings of the Eighth International Conference on Very Large Data Bases*, (New York: IEEE), pp. 165–180.

Kleiner, A., 1989, Storage methods for fast access to large cartographic data collections: an empirical study. In *Proceedings of Auto Carto 9*, (Falls Church: ACSM), pp. 416–435.

Klinger, A. and Dyer, C. R., 1976, Experiments in picture representation using regular decomposition. *Computer Graphics and Image Processing*, **5**, pp. 68–105.

Klopprogge, M. R., 1981, TERM: an approach to include the time dimension in the entity-relationship model. In *Entity-Relationship Approach to Information Modeling and Analysis*, edited by P.P.S. Chen (Sangers, CA: ER Institute), pp. 477–512.

Klopprogge, M. R. and Lockermann, P. C., 1983, Modeling information preserving databases: consequences of the concept of time. In *Proceedings of the Ninth International Conference on Very Large Data Bases*, (New York: IEEE), pp. 399–416.

Knott, G. O., 1975, Hashing functions. *The Computer Journal*, **18**, pp. 265–278.

Knuth, D. E., 1971, Mathematical analysis of algorithms. Stanford University Computer Science Department. Technical Report 71–206.

Knuth, D. E., 1973, *The Art of Computer Programming, Volume 3: Sorting and Searching*. (Reading, Massachusetts: Addison-Wesley).

Kriegel, H. P. and Seeger, B., 1988, PLOP-hashing: a grid file without directory. In *Proceedings of the International Conference on Data Engineering*, (New York: IEEE), pp. 369–376.

Kung, C. H., 1985, On verification of database temporal constraints. In *Proceedings of the SIGMOD '85 Conference*, (New York: ACM), pp. 169–179.

Langran, G., 1988, Temporal GIS design tradeoffs. In *Proceedings of GIS/LIS '88* Volume 2, (Falls Church, VA: ACSM), pp. 890–899.

Langran, G., 1989a, Accessing spatiotemporal information in temporal GIS. In *Proceedings of Auto-Carto 9*, (Falls Church, VA: ACSM), pp. 191–198.

Langran, G., 1989b, A review of temporal database research and its use in GIS applications. *International Journal of GIS*, **3** (3), pp. 215–232.

Langran, G., 1990, Tracing temporal information in an automated nautical charting system. *Cartography and GIS*, **17**, (4), pp. 291–299.

Langran, G. E. and Poiker, T. K., 1986, Integration of names selection and names placement. In *Proceedings of the Second International Symposium on Spatial Data Handling*, (Amherst, NY: IGU), pp. 50–64.

Langran, G. and Clawson, M., 1986, Planning for a naval electronic chart. Naval Ocean Research and Development Activity, NR 119, Bay St. Louis, Mississippi.

Langran, G. and Chrisman, N. R., 1988, A framework for temporal geographic information. *Cartographica*, **25** (3), pp. 1–14.

Lee, D. T. and Wong, C. K., 1980, Quintary trees: a file structure for multi-dimensinal database systems. *ACM Transactions on Database Systems*, **5** (3) pp. 339–353.

Lehan, T. J., 1986, The influence of spatial information on data queries. In *Proceedings of the ASPRS-ACSM Annual Convention*, (Falls Church, VA: ASPRS-ACSM), pp. 250–257.

Levy, M., Pollack, H. and Pomeroy, P., 1970, Motion picture of the seismicity of the earth, 1961–1967. *Bulletin of the Seismological Society of America*, **60** (3), pp. 1015–1016.

Lindsay, B., Haas, L., Mohan, C., Pirahesh, H. and Wilms, P., 1986, A snapshot differential refresh algorithm. In *Proceedings of the SIGMOD '86 Conference*, (New York: ACM) pp. 53–60.

Liou, J. H. and Yao, S. B., 1977, Multidimensional clustering for database organizations. *Information Systems*, **2** (4), pp. 187–198.

Lodwick, G. D. and Feuchtwanger, M., 1987, Land-related information systems. Department of Surveying Engineering, University of Calgary, Alberta, UCSE Report Number 10010.

Lorie, R. A., and Meier, A., 1984, Using a relational database for geographic databases. *Geoprocessing*, **2**, pp. 243–257.

Lukatela, H., 1987, Hipparchus geopositioning model: an overview. In *Proceedings of Auto-Carto 8*, Baltimore, (Falls Church, VA: ACSM), pp. 87–96.

Ludwig, D., Aronson, D. G. and Weinberger, H. F., 1979, Spatial patterning of the spruce budworm. *Journal of Mathematical Biology*, **8**, pp. 217–258.

Lum,. U. Y., Yuen, P. S. T. and Dodd, M. 1971, Key-to-address transform techniques: a fundamental performance study on large existing formatted files. *Communications of the ACM*, **14**, p. 228.

Lum, V., Dadum, P., Erbe, R., Guenauer, J., Pistor, P., Walch, G., Werner, H. and Woodfill, J., 1984, Designing DBMS support for the temporal dimension. In *Proceedings of the SIGMOD '84 Conference*, (New York: ACM), pp. 115–126.

Lynch, K., 1960, *The Image of the City*. (Cambridge, Massachusetts: MIT Press).

Maffini, G. and Saxton, W., 1987, Deriving values from analysis and modification of spatial data. In *Proceedings of the International GIS Symposium*, Volume 3, (Washington DC: NASA), pp. 286–288.

Mandelbaum, S. J., 1984, Temporal conventions in planning discourse. *Environment and Planning B*, **11**, pp. 5–13.

Mandelbrot, B. B., 1977, *The Fractal Geometry of Nature*. (San Francisco: W. H. Freeman).

Manning, H., (ed.), 1910, *The Fourth Dimension Simply Explained*. (New York: Munn).

Mark, D. M., 1979, Phenomenon-based data structures and digital terrain models. *Geo-processing*, **1**, 27.

Mark, D. M. and Aronson, P. A., 1984, Scale-dependent fractal dimensions of topographic surfaces. *Mathematical Geology*, **16**, pp. 671–683.

Mark, D. M. and Lauzon, J. P., 1985, Approaches for quadtree-based geographic information systems at continental and global scales. In *Proceedings of Auto-Carto 7*, (Falls Church, VA: ACSM), pp. 355–364.

Mark, D. M. and Cebrian, J. A. 1986, Oct-trees: a useful data structure for the processing of topographic and sub-surface data. In *Proceedings of the ASPRS-ACSM Annual Meeting*, Volume 1, (Falls Church, VA: ASPRS-ACSM), pp. 104–113.

Maurer, W. D., 1968, An improved hash code for scatter storage. *Communications of the ACM*, **11**, 40.

McCormick, B. W., 1987, Visualization in scientific computing. *ACM SIGGRAPH*, **21**, (6), pp. 21–29.

McDermott, D., 1982, A temporal logic for reasoning about processes and plans. *Cognitive Science*, **6**, pp. 101–155.

McKenzie, E., 1986, Bibliography: temporal databases. *SIGMOD Record*, **15** (4), pp. 40–52.

McKenzie, E. and Snodgrass, R., 1987, Extending the relational algebra to support transaction time. In *Proceedings of the SIGMOD '87 Conference*, (New York: ACM), pp. 467–478.

McNeil, D. R., 1877, *Interactive Data Analysis*. (New York: John Wiley and Sons).

Meredith, P., 1972, The psychophysical structure of temporal information. In *The Study of Time: Proceedings of the First Conference of the International Society for the Study of Time*. (New York: Springer-Verlag), pp. 259–273.

Moellering, H., 1973, The potential uses of a computer animated film in the analysis of geographical patterns of traffic crashes. *Accident Analysis and Prevention*, **8**, pp. 215–227.

Moellering, H., 1980a, Strategies of real-time cartography. *The Cartographic Journal*, **17**, pp. 12–15.

Moellering, H., 1980b, The real-time animation of three-dimensional maps. *The American Cartographer*, **7**, pp. 67–75.

Moellering, H. (ed.), 1987, A draft proposed standard for digital cartographic data. USGS Open-File Report pp. 87–308.

Morehouse, S., 1985, ARC/INFO: a geo-relational model for spatial information. In *Proceedings of Auto-Carto 7*, (Falls Church, VA: ACSM), pp. 388–397.

Morrill, R. L., 1963, The development of spatial distributions of towns in Sweden: an historical-predictive approach. *Annals of the Association of American Geographers*, **53**, pp. 1–14.

Morrill, R. L., 1977, Efficiency and equity of optimal location models. *Geographic Analysis*, **9**, pp. 215–226.

Morrill, R. L., 1982, Trends in trade. *Growth and Change*, **13**, pp. 46–49.

Morris, R., 1968, Scatter storage techniques. *Communciations of the ACM*, **11**, pp. 38–44.

Morris, R., 1984, *Times Arrows: Scientific Attitudes Toward Time*. (New York: Simon and Schuster).

Morrison, J., 1974, A theoretical framework for cartographic generalization with emphasis on the process of symbolization. *International Yearbook of Cartography*, **14**, pp. 115–127.

Morrison, J., (ed,) 1988, The proposed standard for digital cartographic data. *The American Cartographer*, **15** (January), pp. 142.

Morton, G. M., 1966, A computer oriented geodetic data base, and a new technique in file sequencing. IBM Canada Ltd, Ottawa.

Muehrcke, P. C., 1978, *Map Use*. (Madison, Wisconsin: JP Publications).

Naps, T. L. and Singh, B., 1986, *Introduction to Data Structures using Pascal*. (St. Paul, Minnesota: West).

National Ocean Service, 1985, *Desk Reference Guide*. (NOS: Rockville, Maryland).

Newton-Smith, W. H., 1980, *The Structure of Time*. (Boston: Routledge and Kegan Paul).

Nievergelt, J., Hinterberger, H. and Sevcik, K. C., 1984, The grid file: an adaptable, symmetric multikey file structure. *ACSM Transactions on Database Systems*, **9** (1), pp. 38–71.

Noronha, V., 1988, A survey of hierarchical partitioning methods for vector images. In *Proceedings of the Third International Symposium on Spatial Data Handling*, (Amherst, NY: IGU), pp. 185–200.

Nyerges, T., 1980a, Representing spatial properties in cartographic data bases. In *Proceedings of the ACSM Springs Meetings*, St. Louis, ACSM, pp. 29–41.

Nyerges, T., 1980b, *Modeling the structure of cartographic information for query processing*. PhD Dissertation, Ohio State University, Columbus, (Unpublished).

Nyerges, T., 1989, Graphical methods to support micro-macro spatial modeling in geographical information system. In *Proceedings of the International Cartographic Association, (unpublished paper)*.

Ordnance Survey, 1981, Report on the study of revision. Southampton.

Orenstein, J. A., 1982, Multidimensional TRIEs used for associative searching. *Information Processing Letters*, **14** (4), pp. 150–157.

Orenstein, J. A., 1986, Spatial query processing in an object-oriented database system. In *Proceedings of the SIGMOD '86 Conference*, (New York: pp. 172–181).

Overmyer, R. and Stonebraker, M., 1982, Implementation of a time expert in a database system. *SIGMOD Record*, **12** (3), pp. 51–59.

Parkes, D. and Thrift, N., 1980, *Times, Spaces and Places*. (New York: John Wiley and Sons).

Perkal, J., 1956, On epsilon length *Bulletin de L'Academie Polonaise des Sciences*, **4**, pp. 399–403.

Peterson, J. L. and Abraham, S., 1985, *Operating System Concepts*. (Reading, Massachusetts: Addison-Wesley).

Peterson, W. W., 1957, Addressing for random-access storage. *IBM Journal of Research and Development*, **1**, pp. 130–146.

Peuquet D. J., 1984, A conceptual framework and comparison of spatial data models. *Cartographica*, **21** (4), pp. 66–113.

Peuquet, D. J., 1988, Representations of geographic space: toward a conceptual synthesis. *Annals of the Association of American Geographers*, **78**, pp. 375–394.

Pfaltz, J. L., 1978, The costs of data access. In *Harvard Papers on GIS*, Volume 6, edited by G. Dutton, (Reading, Massachusetts: Addison-Welsey).

Poiker, T. K., 1976, A theory of the cartographic line. *International Yearbook of Cartography*, **16**, pp. 134–143.

Poiker, T. K., 1979, Computer cartography and the structure of its algorithms, *World Cartography*, **15**, pp. 71–79.

Poiker, T. K. and Chrisman, N., 1975, Cartographic data structures. *The American Cartographer*, **2**, pp. 55–69.

Pooch, U. W. and Nieder, A., 1973, A survey of indexing techniques for sparse matrices. *ACM Computing Surveys*, **5**, pp. 109–133.

Pred, A., 1977, The choreography of existence: comments on Hagerstrand's time-geography and its usefulness. *Economic Geography*, **52**, pp. 207–221.

Price, C., 1971, Table lookup techniques. *ACM Computing Surveys*, **3** (2), pp. 49–65.

Price, S., 1989, Modelling the temporal element in LISs. *International Journal of GIS*, **3** (3), pp. 233–243.

Prigogine, I., 1985, Time and human knowledge. *Environment and Planning B*, **12**, pp. 5–20.

Rathmann, P., 1984, Dynamic data structures on optical disks. In *Proceedings of the International Conference on Data Engineering*, (New York: IEEE), pp. 175–180.

Reddy, D. R. and Rubin, S., 1978, Representation of three-dimensional objects. Report CMU-CS-78-113. Department of Computer Science, Carnegie-Mellon.

Rescher, N. and Urquhart, A., 1971, *Temporal Logic*. (New York: Springer-Verlag).

Rhind, D. W., 1973, Generalisation and realism within automated cartographic systems. *The Canadian Cartographer*, **10** (1), pp. 51–62.

Rhind, D. W. and Green, N. P. A., 1988, Design of a geographic information system for a heterogeneous scientific community. *International Journal of GIS*, **2**, pp. 171–189.

Rhind, D., Adams, T., Fraser, S. E. G. and Elston, M., 1983, Towards a national digital topographic data base: experiments in mass digitising, parallel processing, and detection of change. In *Proceedings of Auto-Carto 6*, Volume 2, (Ottawa: Steering Committee of the Sixth International Symposium on Automated Cartography).

Robinson, A. H. and Petchenik, B. B., 1976, *The Nature of Maps*, (Chicago and London: University of Chicago Press).

Robinson, J. T., 1981, The K-D-B tree: a search structure for large multidimensional dynamic indexes. In *Proceedings of the SIGMOD '81 Conference*, (New York: ACM), pp. 186–191.

Rosenblat, S., 1980, Population models in a periodically fluctuating environment. *Journal of Mathematical Biology*, **9**, pp. 23–36.

Rosenzweig, S., 1971, Paradox of enrichment: destabilization of exploitation ecosystems in ecological time. *Science*, **171**, pp. 385–387.

Ross, J., 1985, Detecting land use change on Omaha's urban fringe using a geographic information system. In *Proceedings of Auto-Carto 7*, (Falls Church, VA: ACSM), pp. 463–471.

Rotem, D. and Segev, A., 1987, Physical organization of temporal data. In *Proceedings of the International Conference on Data Engineering*, (New York: IEEE), pp. 547–553.

Rothnie, J. B. and Lozano, T., 1974, Attribute-based file organization in a paged environment. *Communications of the ACM*, **17**, pp. 63–69.

Roussopoulos, N. and Leifker, D., 1985, Direct spatial search on pictorial databases using packed R-trees. In *Proceedings of SIGMOD '85*, (New York: ACM), pp. 17–31.

Rucker, R. von B., 1977, *Geometry, Relativity, and the Fourth Dimension*. (New York: Dover Books).

Rucker, R., 1984, *The Fourth Dimension: Toward a Geometry of Higher Reality*. (New York: Houghton Mifflin).

Salton, G. and Wong, A., 1978, Generation and search of clustered files. *ACM Transaction on Database Systems*, **3** (4), pp. 321–346.

Samet, H., 1984, The quadtree and related hierarchical data structures. *ACM Computing Surveys*, **16** (2), pp. 220–248.

Samet, H. and Webber, R. E., 1985, Storing a collection of polygons using quadtrees. *ACM Transactions on Graphics*, **4** (3), pp. 182–222.

Samet, H., Rosenfeld, A., Shaffer, C. A. and Webber, R. E., 1984, Use of hierarchical data structures in geographic information systems. In *Proceedings of the First International Symposium on Spatial Data Handling*, Volume 2, (Amherst, NY: IGU), p. 412–430.

Samsel, T. B. and Colten, C. E., 1988, Surface hydrology and the accumulation of hazardous materials on the American Bottoms, 1890–1980. In *Proceedings of GIS/LIS '88*, Volume 2, (Falls Church, VA: ACSM), pp. 861–866.

Samson, P. J. and Small, M. J., 1984, Atmospheric trajectory models for diagnosing the sources of acid precipitation. In *Modeling of Total Acid Precipitation Impacts*, edited by J. L. Schnoor. (Boston: Buttenworth).

Sarnak, N. and Tarjan, R. E., 1986, Planar point position using persistent search trees. *Communications of the ACM*, **29**, pp. 669–679.

Schiel, U., 1983, An abstract introduction to the temporal-hierarchic data model. In *Proceedings of the Ninth International Conference on Very Large Data Bases*, (New York: IEEE), pp. 322–330.

Schneider, J. B., 1979, Congestion displays: a policy oriented feasibility assessment. Report 79-1, UMSTA—WA-11-0002, Seattle, Washington.

Schneiderman, B., 1974, A model for optimizing indexed file structures. *International Journal of Computer and Information Science*, **3** (1), pp. 93–103.

Schubert, L. K., Papalaskaris, M. A. and Taugher, J., 1983, Determining type, part, color, and time relationships. *Computer*, **16**, pp. 53–60.

Sedgewick, R., 1983, *Algorithms*. (Reading, Massachusetts: Addison-Wesley).

Segev, A. and Shoshani, A., 1987, Logical modeling of temporal data. In *Proceedings of the SIGMOD '87 Conference*, (New York: ACM), pp. 454–466.

Sernadas, A., 1980, Temporal aspects of logical procedure definition. *Information Systems*, **5** (3), pp. 167–187.

Seymour, G. A., Utano, J. J. and Marquette, J. F., 1981, Automated approach to space-time statistical mapping. In *Computer Mapping Applications in Urban, State, and Federal Government*, Edited by P. A. Moore, (Cambridge: Harvard Graduate School of Design), pp. 36–49.

Seymour, W., (ed), 1980, *A History of the Ordnance Survey*. (Folkestone, Kent: William Dawson and Sons).

Shapiro, L. and Haralick, R. M., 1980, A spatial data structure. *Geo-Processing*, **1**, 313–337.

Sherman, J. C. and Tobler, W. R., 1957, Multiple use concept in cartography. *The Professional Geographer*, **9** (5), pp. 5–7.

Sinton, D., 1978, The inherent structure of information as a constraint to analysis: mapped thematic data as a case study. In *Harvard Papers on GIS*, Volume 7, edited by G. Dutton, (Reading, Massachusetts: Addison-Wesley).

Six, H. W. and Widmayer, P., 1988, Spatial searching in geometric databases. In *Proceedings of the International Conference on Data Engineering*, (New York: IEEE), pp. 496–503.

Smith, J. M. and Smith, D. C. P., 1977, Data base abstraction: aggregation. *Communications of the ACM*, **20**, pp. 405–413.

Smith, T. R., Menon, S., Star, J. L. and Estes, J. E., 1987, Requirements and principles for the implementation and construction of large-scale geographic information systems. *International Journal of GIS*, **1** (1), pp. 13–31.

Snodgrass, R., (ed.), 1986, Research concerning time in databases: project summaries. *SIGMOD Record*, **15** (4), pp. 19–39.

Snodgrass, R., 1987, The temporal query language TQuel. *ACM Transactions on Database Systems*, **12**, pp. 247–298.

Snodgrass, R. and Ahn, I., 1985, A taxonomy of time in databases. In *Proceedings of the SIGMOD '85 Conference*, (New York: ACM), pp. 236–245.

Snodgrasss, R. and Ahn, I., 1986, Temporal databases. *Computer*, **19**, pp. 35–42.

Srnka, E., 1970, The analytical solution of regular generalization in cartography. *International Yearbook of Cartography*, **10**, pp. 48–61.

Steinitz, C., Parker, P. and Jordan, L., 1976, Hand-drawn overlays: their history and prospective use. *Landscape Architecture*, **66** (5), pp. 444–455.

Stenhouse, H., 1979, Selection of towns on derived maps. *The Cartographic Journal*, **16**, pp. 30–39.

Stonebraker, M., 1986, Triggers and inference in database systems. In *On Knowledge Base Management Systems*, edited by M. L. Brodie and J. Mylopoulos, (New York: Springer-Verlag), pp. 297–314.

Stonebraker, M., 1988, Future trends in database systems. In *Proceedings of the International Conference on Data Engineering*, (New York: IEEE), pp. 222–231.

Studer, R., 1986, Modeling time aspects of information systems. In *Proceedings of the International Conference on Data Engineering*, (New York: IEEE), pp. 364–372.

Sukhov, V. I., 1970, Applications of information theory in the generalization of map contents. *International Yearbook of Cartography*, **10**, pp. 41–47.

Sundgren, B., 1975, *Theory of Data Bases*. (New York: Mason/Charter).

Szegö, J., 1987, *Human Cartography: Mapping the World of Man*. (Stockholm: Swedish Council for Building Research).

Tamminen, M., 1981, The EXCEL method for efficient geometric access to data. Mathematics and Computer Science Series, 34, (Helsinki: Acta Polytechnica Scandinavia).

Tamminen, M., 1982, the extendible cell method for closest point problems. *BIT*, **22**, 27–41.

Tansel, A. U., 1987, A statistical interface for historical relational databases. In *Proceedings of the International Conference on Data Engineering*, (New York: IEEE), pp. 538–546.

Tenenbaum, A. M. and Augenstein, M. J., 1981, *Data Structures Using Pascal*. (Englewood Cliffs, New Jersey: Prentice-Hall).

Thompson, M. M., 1979, *Maps for America*. (U.S. Geological Survey: Government Printing Office).

Thrift, N., 1977, *An Introduction to Time Geography*. (Norwich: Geo-Abstracts).

Tobler, W., 1964, An experiment in the computer generalization of maps. Department of Geography, University of Michigan. Technical Report 1, December.

Tobler, W., 1970a, Geographical filters and their inverses. *Geographical Analysis*, 1, pp. 234–253.

Tobler, W., 1970b, A computer movie simulating urban growth in the Detroit region. *Economic Geography*, **46**, pp. 234–240.

Tobler, W., 1979, Cellular geography. In *Philosophy in Geography*, edited by S. Gaile and G. Olsson. (Dordrecht, Holland: D. Reidel), pp. 379–386.

Topfer, F. and Pillewizer, W., 1966, The principles of selection. *The Cartographic Journal*, 3, pp. 10–16.

Tomlinson, R. and Boyle, A. R., 1981, the state of development of systems for handling natural resources inventory data. *Cartographica*, 18 (4), pp. 65–95.

Tomlinson, R. F., Calkins, H. W. and Marble, D. F., 1976, Computer handling of geographic data. (Paris: UNESCO Press).

Tsichritzis, D. C. and Lochovsky, F. H., 1982, *Data Models*. (Englewood Cliffs, New Jersey: Prentice-Hall).

Tufte, E. R., 1983, *The Visual Display of Quantitative Information*. (Cheshire, Connecticut: Graphics Press).

Tukey, J.W., 1977, *Exploratory Data Analysis*. (Reading, Massachusetts: Addison-Welsey Publishers).

Tuori, M. and Moon, G. C., 1984, A topographic map conceptual model. In *Proceedings of the First International Symposium on Spatial Data Handling*, Volume 1, (Amherst IGU), pp. 28–37.

Turner, S., 1986, Dynamic cartography in a prototype satellite monitoring system. In *Proceedings of the ACSM-ASPRS Annual Convention*, Volume 1, (Falls Church, VA: ACSM-ASPRS), pp. 222–227.

Urban, S. D. and Delcambre, L. M. L., 1986, An analaysis of the structural, dynamic and temporal aspects of semantic data models. In *Proceedings of the International Conference on Data Engineering*, (New York: IEEE), pp. 382–389.

Van Oosterom, P., 1989, A reactive data structure for geographic information systems. In *Proceedings of Auto-Carto 9*, (Falls Church, VA: ACSM), pp. 665–674.

Van Roessel, J., 1986, Design of a spatial data structure using the relational normal forms. In *Proceedings of the Second International Symposium on Spatial Data Handling*, (Amherst: IGU), pp. 251–272.

Veen, A. H., 1986, Dataflow machine architecture. *ACM Computing Surveys*, 18, pp. 365–396.

Velleman, P. F. and Hoaglin, D. C., 1981, *Applications, Basics, and Computer Methods of Explatory Data Analysis*. (Boston: Duxbury Press).

Vrana, R., 1989, Historical data as an explicit component of land information systems. *International Journal of GIS*, 3 (1), pp. 33–49.

Walsh, S. J., Lightfoot, D. R. and Butler, D. R., 1987, Assessment of inherent and operational errors in geographic information systems. In *Proceedings of the Annual ASPRS-ACSM Convention*, Volume 5, (Falls Church, VA: ASPRS-ACSM), pp. 24–35.

Wang, F. and Newkirk, R., 1987, A GIS-supported digital remote sensing land cover change detection system. *Proceedings of the ASPRS-ACSM Annual Convention*, Volume 6, (Falls Church, VA: ASPRS-ACSM), pp. 53–62.

Wasowski, R. J. and Ferretti, S. B., 1987, A time-lapse analysis of the Mississippi Delta using Landsat MSS Band 4 IR2 photographic imagery. In *Proceedings of the ASPRS-ACSM Annual Convention*, Volume 1, (Falls Church, VA: ASPRS-ACSM), pp. 386–392.

Waugh, T. C., 1986, A response to recent papers and articles on the use of quadtrees for geographic information systems. In *Proceedings of the Second International Symposium on Spatial Data Handling*, (Amherst: IGU), pp. 33–37.

Waugh, T. C. and Healey, R. G., 1986, The GEOVIEW design: a relational database approach to geographic data handling. In *Proceedings of the Second International Symposium on Spatial Data Handling*, (Amherst: IGU), pp. 193–212.

White, M., 1975, Map editing using a topological access system. In *Proceedings of Auto-Carto 2*, (Falls Church, VA: ACSM), pp. 422–429.

White, M., 1978, The cost of topological file access. In *Harvard Papers on GIS*, Volume 6, edited by G. Dutton. (Reading, Massachusetts: Addison Wesley).

White, M., 1983, Tribulations of automated cartography and how mathematics helps. In *Proceedings of Auto-Carto 6*, Volume 1, (Ottawa: Steering Committee of the Sixth International symposium on Automated Cartography).

White, M., 1987, Digital map requirements of vehicle navigation. In *Proceedings of Auto-Carto 8*, (Falls Church, VA: ACSM), pp. 552–561.

Whitelaw, J. S., 1972, Scale and urban migration behavior. *Australian Geographic Studies*, **10**, pp. 101–106.

Whittlesey, D., 1945, The horizon of geography. *Annals of the Association of American Geographers*, **35**, pp. 1–36.

Wiede, B., 1977, A survey of analysis techniques for discrete algorithms. *ACM Computing Surveys*, **9** (4), pp. 291–313.

Williams, J. W. J., 1964, Algorithm 232 Heapsort. *Communications of the ACM* **7**, pp. 347–348.

Wirth, N., 1976, *Algorithms + data structures = programs*. (Englewood Cliffs, New Jersey: Prentice-Hall).

Wood, D. and Fels, J., 1986, Designs on signs: myth and meaning in maps. *Cartographica*, **23**, pp. 54–103.

Youngmann, C. S., 1978, A linguistic approach to map description. In *Harvard Papers on GIS*, Volume 7, edited by G. Dutton. (Reading, Massachusetts: Addison Wesley).

Yuval, G., 1975, Finding near neighbors in *k*-dimensional space. *Information Processing Letters*, **3** (4), pp. 113–114.

Zhu, M., Loh, N. M. K. and Siy, P., 1987, Towards the minimum set of primitive relations in temporal logic. *Information Processing Letters*, **26**, pp. 121–126.

Index